CHEMISTRY RESEARCH AND APPLICATIONS

FLOW THROUGH OPTOSENSORS

CHEMISTRY RESEARCH AND APPLICATIONS

Additional books in this series can be found on Nova's website under the Series tab.

Additional E-books in this series can be found on Nova's website under the E-book tab.

CHEMISTRY RESEARCH AND APPLICATIONS

FLOW THROUGH OPTOSENSORS

ANTONIO RUIZ MEDINA

Nova Science Publishers, Inc.
New York

Copyright © 2012 by Nova Science Publishers, Inc.

All rights reserved. No part of this book may be reproduced, stored in a retrieval system or transmitted in any form or by any means: electronic, electrostatic, magnetic, tape, mechanical photocopying, recording or otherwise without the written permission of the Publisher.

For permission to use material from this book please contact us:
Telephone 631-231-7269; Fax 631-231-8175
Web Site: http://www.novapublishers.com

NOTICE TO THE READER

The Publisher has taken reasonable care in the preparation of this book, but makes no expressed or implied warranty of any kind and assumes no responsibility for any errors or omissions. No liability is assumed for incidental or consequential damages in connection with or arising out of information contained in this book. The Publisher shall not be liable for any special, consequential, or exemplary damages resulting, in whole or in part, from the readers' use of, or reliance upon, this material. Any parts of this book based on government reports are so indicated and copyright is claimed for those parts to the extent applicable to compilations of such works.

Independent verification should be sought for any data, advice or recommendations contained in this book. In addition, no responsibility is assumed by the publisher for any injury and/or damage to persons or property arising from any methods, products, instructions, ideas or otherwise contained in this publication.

This publication is designed to provide accurate and authoritative information with regard to the subject matter covered herein. It is sold with the clear understanding that the Publisher is not engaged in rendering legal or any other professional services. If legal or any other expert assistance is required, the services of a competent person should be sought. FROM A DECLARATION OF PARTICIPANTS JOINTLY ADOPTED BY A COMMITTEE OF THE AMERICAN BAR ASSOCIATION AND A COMMITTEE OF PUBLISHERS.

LIBRARY OF CONGRESS CATALOGING-IN-PUBLICATION DATA

Ruiz, Antonio, 1971-
 Flow through optosensors / Antonio Ruiz.
    p. cm.
 Includes bibliographical references and index.
 ISBN 978-1-61942-467-8 (soft cover)
 1. Flow injection analysis. 2. Solid-phase analysis. 3. Optical spectroscopy. I. Title.
 QP519.9.F55R85 2011
 543'.54--dc23

2011048502

Published by Nova Science Publishers, Inc. † *New York*

CONTENTS

Abstract		**vii**
Chapter 1	Introduction	**1**
Chapter 2	Solid Phase Specroscopy	**5**
	2.1. Theoretical Fundamentals	5
	2.2. SPS and FTO Methodologies	7
	2.3. Types of Solid Support	10
	2.4. Measurement of the Analyte and Regeneration of the Solid Support	14
Chapter 3	Bead Injection Spectroscopy	**17**
Chapter 4	Flow-Through Cells	**23**
Chapter 5	Flow Methodologies in FTO	**27**
	5.1. Flow Injection Analysis	31
	5.2. Sequential Injection Analysis	38
	5.3. Multicommutation	42
	5.3.1. MCFIA	43
	5.3.2. MPFS	46
	5.3.3. MSFIA	47
	5.4. Comparison of Flow Methodologies	48

Chapter 6	Classification of FTO in Terms of Number of Analytes	51
	6.1. Mono-Sensing Optosensors	51
	6.2. Multi-Sensing Optosensors	52
	6.2.1. Use of two different sensing zones	52
	6.2.2. Separation in a minicolumn	53
	6.2.3. Separation in the flow-cell	54
	6.2.4. Separation with mathematical treatment	55
Chapter 7	Detection Techniques	57
	7.1. UV-Visible Spectroscopy	57
	7.2. Luminescence	58
	7.2.1. Fluorescence	59
	7.2.2. Phosphorescence	61
	7.2.3. Chemiluminiscence	61
	7.2.4. Lanthanide-sensitized luminescence	63
	7.3. Reflectometry	66
	7.4. Vibrational Spectroscopy	67
Chapter 8	Applications	71
	8.1. Pharmaceuticals and Biological Samples	71
	8.2. Food	87
	8.3. Trace elements	89
	8.4. Pesticides	91
Chapter 9	Conclusions and Trends	97
References		99

ABSTRACT

This book focuses on the main characteristics of direct solid-phase measurements in flow systems. The fundamentals and the technical status of flow-through optosensing together with the main detection techniques are described, including the discussion of advantages and limitations and practical guidelines to the successful implementation of this approach. In order to give readers insight into the features and the potential of flow through optosensors, relevant examples of strategies and types of configuration are studied. The use of this methodology in flow-based systems is advantageous in view of the reproducible handling of solutions in retention and eluting steps of the analytes. Moreover, flow through optosensors can be exploited to increase sensitivity, minimize reagent consumption as well as waste generation, and improve selectivity for simultaneous determination based on selective retention or differences in sorption rates of the analytes.

A description of concepts and applications of each flow technique such as flow injection analysis, sequential injection analysis and multicommuted flow analysis are reviewed. Main configurations of these approaches are highlighted, paying special attention to both the recent incorporation of new detection techniques in optosensing and the design of multiparameter sensors. The reports cited show the characteristics of these systems including easy automation and low cost of the equipment. Selected applications in diverse fields, such as pharmaceutical, food, and environmental analysis are discussed.

Chapter 1

INTRODUCTION

The combined use of an active solid support to pre-concentrate the analyte (or its reaction product) with the direct measurement of the light absorption of the species of interest sorbed on the solid phase was proposed by Yoshimura [1] in 1976. This methodology, first called ion-exchanger colorimetry (as the first supports used were ion-exchangers) is nowadays more appropriately called Solid Phase Spectroscopy (SPS) due to the employment of different spectroscopic detections. Since then, many reports have been published concerning this methodology. The most remarkable features of SPS are its high sensitivity and selectivity. SPS methods are more sensitive than the conventional corresponding solution methods, due to the concentration of the target analyte and the direct light absorption or emission measurement on a solid-phase, which are simultaneously carried out without the elution of the target compound. Sensitivity can be easily increased one hundred-fold without the employment of any expensive apparatus. On the other hand, the selectivity can also be highly improved when the interaction between the target chemical species and the solid phase beads is quite different from that one of co-existing species.

Two different general procedures have been employed in SPS, i.e. batch and flow methodologies. In batch mode, due to a fairly large background attenuance of the solid phase (surface reflection and/or a diffuse reflection of the solid layer) sensitivity enhancement by making the light path of the solid particle layer longer is difficult unless large volumes of sample are used (normally, measurements are carried out within 1-2 mm light path length). Moreover, batch method is slightly time-consuming and requires both bead manipulation stages to pack the solid particles into a cell (e.g., load, unload, filtration) and use of large amounts of solid material. In the second procedure, the flow method, solid phase retention is implemented with on line spectroscopic detection so originating Flow Injection-Solid Phase Spectroscopy (FI-SPS) methodology [2]. In this way, the advantages of SPS were added to those ones intrinsic from flow methods (rapidity, commodity, automation, less consumption of reagents and solid support, etc). In these FI-SPS systems the (micro)zone of the solid phase, where the signal is continuously monitored, is surrounded by a continuous stream flowing through it. The separation and retention of the species of interest on the solid phase takes place in the detection area itself and simultaneously with it. The sample plug is inserted in the stream and so, the radiation directly interacts with the solid surface integrated in the detection area. This principle is called Flow Through Optosensors (FTO) [3,4] which is illustrated in Figure 1.

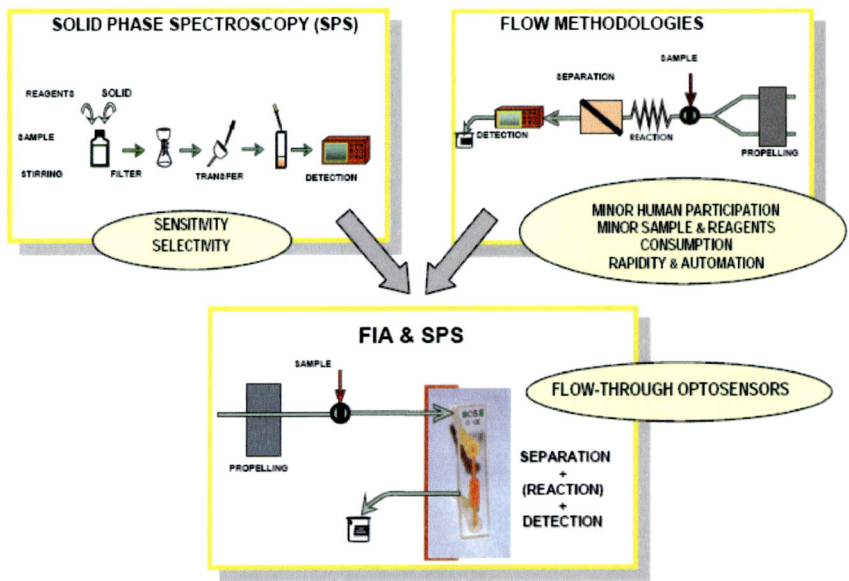

Figure 1. Schematic diagram of FTO.

When a FTO is put in direct contact with the sample, it responds selectively to the species of interest, providing a reversible, rapid and continuous response which is transduced via a non destructive (molecular) spectroscopic detector. This methodology makes it possible to reduce the sample solution volume and simplify the procedures of derivatization of the analyte by filling the flow-cell with the solid particles (compared to batch mode). Improvements in analytical performance, in terms of accuracy, reliability, precision and sample throughput, are also worth mentioning.

Due to the remarkable progress in the recent flow analysis systems the applicability of FTO has been expanded to various chemical species in a great variety of samples. The main applications of FTO were focused firstly towards inorganic analytes although, during the last few years, a high number of FTO have been developed for the determination of organic compounds. Today, research efforts devoted to this development show a very promising research area which produces very simple and inexpensive analytical procedures with remarkable analytical features. In this sense, the main flow methodologies employed in recent years, in addition to conventional Flow Injection Analysis (FIA) [5], have been sequential injection analysis (SIA) [6] and multicommuted flow techniques [7].

In this book relevant examples of strategies and configurations developed in FTO are presented, giving the readers a view of the potential of this methodology when compared to other alternative procedures. On the other hand, it should be enhanced that this development is relevant in the emerging field of green analytical chemistry, due to the reduction of the amount and the toxicity of solvents and reagents used in the steps of sample pre-treatment and measurement as a consequence of automation and miniaturization [8,9].

Chapter 2

SOLID PHASE SPECROSCOPY

2.1. THEORETICAL FUNDAMENTALS

In flow-through sensors the measurement of the analytical signal is directly relationed to the concentration of the analyte in the sample injected. In the case of spectrophotometric sensors, absorbance (really attenuation) on the solid support at the working wavelength, A, consists of several components [10]:

$$A = A_A + A_R + A_S,$$

where A_A is the absorbance of the analyte (or analytical derivative) sorbed on the resin, A_R is the absorbance of the background (solid support plus reagent, if it is the case), and A_S, that of the interstitial solution between the resin beads (which can be neglected as compared with the other terms). The packing of the resin beads in the flow cell affects the values of A_A and A_R. However, when the system is flowing for a few seconds, the packing keeps constant and so the baseline shows a constant value equal to $A_R + A_S \approx A_R$.

Therefore, the analytical peak, A_A, corresponds to the difference between A and $A_R + A_S$. In this way, in the flow-through spectrophotometric system, the analytical signal can be obtained directly by measuring at only one wavelength and successive measurements are performed on the same resin packing. However, in SPS batch mode each measurement is performed on a different resin batch and the packing is also different in each case, so rendering no reproducible measurements when they are performed at one only wavelength. For this reason, measurements at two different wavelengths are needed in SPS in batch mode (one at the absorption maximum of the species of interest, and another at the wavelength where only the resin absorbs light).

The net intrinsic absorbance of the analyte sorbed, A_A, is given by next expression [11] $A_A = \varepsilon_R\, l_R\, C_R$, where ε_R is the apparent molar absorptivity of the analyte in the ion exchanger phase as observed in the flow-through system (Kg mol^{-1} cm^{-1}), l_R the mean light path length through the resin layer which can be supposed equal to 0.1 cm (although it usually will not be exactly this value) and, C_R, the analyte concentration in the solid phase (mol kg^{-1}). When V (L) of a sample at concentration C_o (mol L^{-1}) of analyte is injected in the system, supposing a high value of the distribution ratio (as it is usual), the concentration on the solid phase (C_R, mol Kg^{-1}) will be [12]:

$$C_R = C_o 1000\, V/m_r, \qquad (1)$$

where m_r is the mass of resin (kg) in which the analyte is retained. Therefore:

$$A_A = \varepsilon_R\, l_R\, V\, C_o/m_r, \qquad (2)$$

where m_r is expressed in g. So, keeping V constant, there is a linear relationship between the analytical signal and the initial concentration, C_o, of the analyte in the injected solution, being $\varepsilon_R\, l_R\, V/m_r$ the slope of the calibration line.

An important feature of these flow-through sensing devices is derived from equation (2): keeping C_o constant, A_A increases as V increases. So, a linear relationship between A_A and the injected sample volume can be expected, that is, sensitivity is proportional to the sample volume used for analysis and this can easily be increased just by increasing the injection volume.

In fluorometric sensors, a similar relationship between the analytical signal and the initial concentration of analyte in solution can be established. The fluorescence signal of the analyte sorbed on the solid support, I, is given by the following expression:

$$I = \phi_F\, I_o\, \varepsilon_R\, l_R\, C_R, \qquad (3)$$

where ϕ_F is the fluorescence efficiency in the solid phase, I_o is the intensity of the excitation light beam and, ε_R, l_R and C_R are as above described. From equations (1) and (3) it follows:

$$I = \phi_F I_o \varepsilon_R l_R V C_o/m_r$$

Similar linear relations between the analytical signal and C_o can also be written for phosphorescence and chemiluminescence (CL) sensors.

2.2. SPS AND FTO METHODOLOGIES

Yoshimura et al. [1] described for the first time a photometric procedure based on immobilization of the target species (analyte or a suitable reaction product) on an appropriate solid support (usually micro-beads from a polymeric or a non-polar sorbent material or ion-exchange resins) by establishing an equilibrium between the active sites of the sorbent and the target species in solution. Beads are then collected by filtration and transferred to an appropriate measurement cell as a suspension with a few mL of solution. The analytical spectroscopic property from the target species, typically absorbance or fluorescence, directly retained on the solid support is measured and related to the sample-analyte concentration. This methodology is called SPS.

SPS in batch mode (Figure 2) has been applied to a wide variety of matrices (e.g., waters, foods, soils, plants, pharmaceuticals, and biological fluids) and analytes (cationic and anionic species, organic compounds, etc.). With SPS, traditional elution from the solid support after solid-phase extraction (SPE) preconcentration (with subsequent dilution and partial loss of the preconcentration capabilities achieved with the sorption step) is avoided, achieving higher sensitivity compared with the corresponding conventional homogeneous solution determination.

Interestingly, sensitivity in SPS can be further enhanced by increasing the ratio of sample volume to sorbent mass. The most common procedure typically involves analyte enrichment from a large sample volume (typically 1 L) onto the minimum resin amount required to perform the measurement by filling the optical path of the cell (30-100 mg depending on the instrument light-beam positioning in relation to the cell).

Accordingly, the two main analytical features of SPS methods are high – both sensitivity and selectivity – due to the preconcentration of the species of interest from a relatively large sample volume (typically 100-1500 mL) on a small amount of a solid support, on which detection is performed directly with a non-destructive molecular spectroscopic detector. Very low detection limits were reached for both spectrophotometric and spectrofluorometric procedures. Hence, sensitivity is affected by the value of the ratio between sample volume and amount of solid support. Sensitivity values are several orders of magnitude greater than those from the respective conventional solution methods.

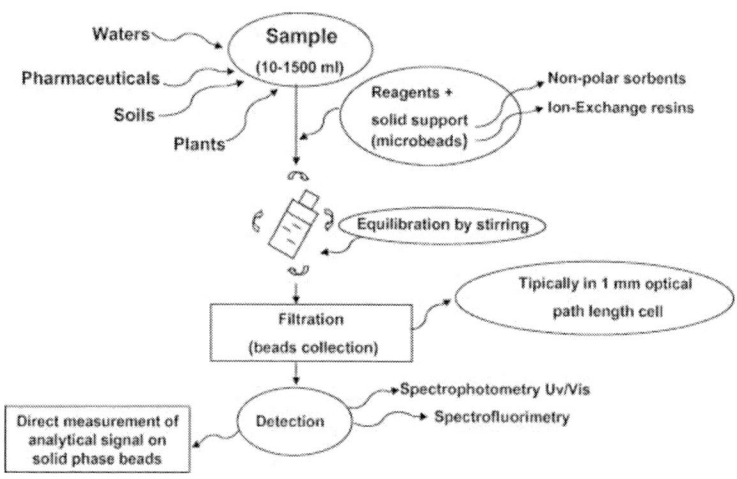

Figure 2. Methodology of SPS in batch mode.

Despite the advantageous features of SPS with respect to the same classical determination in homogeneous solution, it is clear that the batch concept is not easily applicable to routine analysis. It is time-consuming because of reaction development, bead-manipulation stages (e.g., load, unload, filtration) and detection for each determination. Moreover, it requires use of large volumes of sample solution and large amounts of solid material.

One of the more important advances in SPS was achieved when solid-phase detection was implemented with flow-analysis methods by placing solid microbeads inside the flow cell, so performing successive measurements on these microbeads (regenerating the solid phase after each measurement). With this approach, the processes of retention, preconcentration and detection in the analytical procedure are all performed simultaneously and at the same place in the flow system (the flow cell). This approach, shown in Figure 3, is similar to a typical liquid chromatographic process with molecular spectroscopic detection, in which the detector displays the signal (in this case on the column itself) while the chromatographic separation is performed on the separation column (separation and detection occurring in the same place and at the same time, but at low pressure). Figure 3 shows the principle of implementing SPS with flow analysis, including multi-determination approaches.

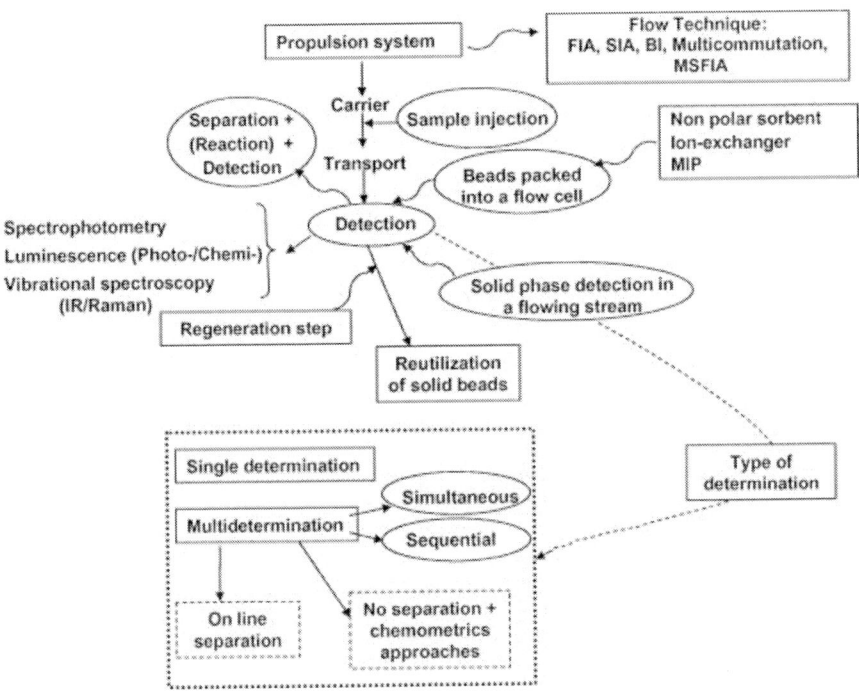

Figure 3. Methodology of SPS in flow mode.

The unique feature of such kinds of FTO, which combine the high selectivity and selectivity shown by SPS with those intrinsic features of flow methods, is miniaturization of SPS with the subsequent reductions in reagent and sample consumption and waste generation. Better accuracy, reliability and precision are also achieved.

A key aspect of FTO that differs from SPS performed in batch mode is the successive reutilization of the solid-phase microbeads placed in the flow cell for a large number of measurements. It involves regeneration of the solid support after each measurement, thus achieving reversible retention of the analyte to get the sensing zone ready for the next determination. This makes automatic flow SPS methodologies more environmental friendly than batch SPS, since a few mg of beads can be reutilized hundreds of times, so dramatically reducing the amount of solid support per determination. In addition, we need to highlight the single instrumentation, short analysis times and low cost per sample, all features making flow-through optosensing one methodology that can be performed in any laboratory and an interesting, cost-effective contributor to green analytical chemistry methodology.

By contrast, we can also outline some drawbacks for these systems. Due to the relatively poor selectivity of spectroscopic measurements, these methods are difficult to apply to trace analysis in complex matrices. However, the main target applications that fit with FTO involve relatively simple matrices (e.g., waters or pharmaceuticals), for which optosensors are more robust.

In addition, sometimes the deactivation of the sensing surface after repeated use is another aspect to be considered in FTO-based systems. The replacement of the solid support in the cell becomes necessary after a certain number of analyses. Moreover, when the monitored species is so strongly retained on the solid sensing beads that the regeneration step is extraordinarily difficult to get, drastically reducing the sensor life time, flow injection renewable surface methodology is recommended. In this one a single-use solid sensing surface is employed, being automatically discarded after each determination, so the use of an eluting solution is not needed.

In general, reagent(s) and sample volumes are much higher in SPS batch mode than SPS flow mode if they are compared. The higher volume of reagent necessary in sensors is only that of the carrier solution. It should be emphasized that the drastic reduction of solid support per sample determination in all sensors is due to its reutilization (up to 400 times lower). In some cases, about 300 or even more determinations can be performed on the same amount of microbeads.

2.3. TYPES OF SOLID SUPPORT

Solid supports are packed in the flow-cell(s) of a conventional non-destructive optical detector. The analytes or their reaction products are immobilized on them temporarily for sensing, so integrating retention and detection, and sometimes reaction.

As it was mentioned before, FTO involve the regeneration of the solid support after each measurement, thus achieving reversible retention of the analyte to get the sensing zone ready for the next determination. A few milligrams of beads can be reutilized hundreds of times. Nevertheless, sometimes the replacement of the solid support in the cell becomes necessary after a certain number of analyses due to the deactivation of the sensing surface after repeated use (e.g., a complex matrix), so reducing its lifetime.

The selection of the solid support strongly depends on the nature of the reagent and/or the corresponding species of target element. This solid support should meet some requirements for the development of a FTO:

- It should be chemically inert to the components of the solutions constituting the flow.
- It should be mechanically resistant to the continuous flow in order to warrant reproducibility in the sensor response.
- The particle size should be large enough to avoid overpressure in the system.
- The retention/elution process should be quick enough.

The most frequently used solid supports in FTO can be divided into three groups: conventional supports, sol-gel supports and molecularly imprinted polymers (MIPs). Conventional supports in turn can be distinguished in three main types: (a) ion-exchange polymers, (b) non-ionic polymeric adsorbents or neutral resins, and (c) non-polar sorbents. For a charged species, an ion-exchange resin or a cross-linked dextran gel-type ion-exchanger are used, while the second and third groups are used for the retention of neutral species.

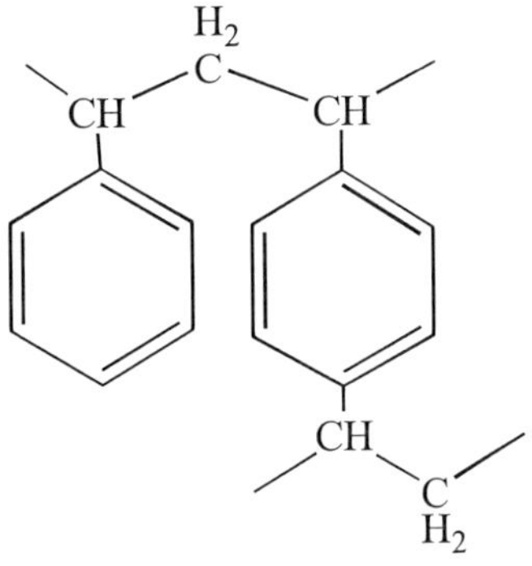

Figure 4. Skeleton of Polystyrene resins.

a) <u>Ion-exchange polymers</u>. Usually, *styrene polymers*, constituted by a hydrophobic aromatic matrix, are discarded when working in UV region because of their very high background in this region. The skeleton of these resins (e.g. Dowex 1 and Dowex 50W) is jointed to different chemical groups (Figure 4).

Dextran polymers, such as Sephadex absorbents, have been the most frequently used (e.g. Sephadex QAE A-25, Sephadex CM C-25 and Sephadex SP C-25), in which different functional groups can be introduced in order to produce cation- or anion- exchanger solid supports. The polysaccharide skeleton of these resins is shown in Figure 5.

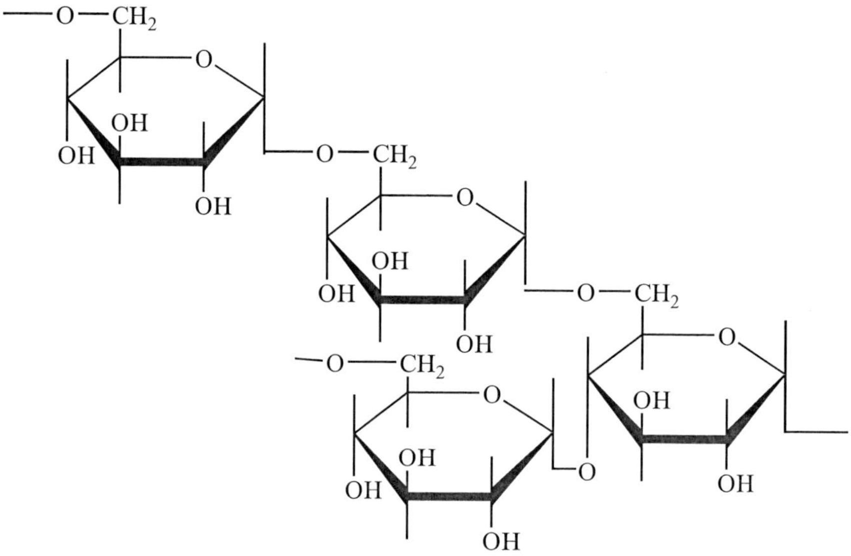

Figure 5. Skeleton of Sephadex resins.

b) <u>Non-ionic polymeric adsorbents or neutral resins</u>. Between these resins, also known like *macroporous polymers*, Amberlite XAD resin is the most used. One of the drawbacks of this type of supports is their high background in the UV and Visible regions. Although many fluorescent impurities due to the aromatic rings in its structure cause a serious background noise in the emission measurement, Amberlite XAD resin has usually been used in luminescence sensors.

c) <u>Non polar sorbents</u>. The employment of a non-polar sorbent such as C_{18} bonded silica beads has allowed many determinations although an important drawback of this material, when compared to the other ones, is its low selectivity because of the adsorptive nature of the retention process.

Sometimes, solid supports packing microcolumns coupled on-line to the flow injection system have also been employed. The use of these microcolumns has two main purposes: (a) the interference removal by use of a support suited to the sorption of the species to be removed and/or (b) to get a multi-determination. The multi-determination can be accomplished by retaining selectively one of the analytes in the microcolumn (placed just before the flow cell) and eluting it later (after developing the signal of another one) with an appropriate reagent [13].

The sol-gel technology provides a simple means to incorporate, at low temperature, organic and biological recognizing elements in a stable inert support. Typically, a sol is first formed by mixing a liquid alcoxide precursor, water, a co-solvent (usually ethanol or methanol) and a catalyst (acid or base) at room temperature. This way, a porous gel network is obtained through continuous monomer hydrolysis and condensation reactions. Afterwards, gel aging and drying can be conducted in order to obtain densified solid matrices. During the steps of hydrolysis, condensation or aging, the recognizing elements can be added and become entrapped in the support net, remaining sterically accessible to small analytes that diffuse into the pore network. Sol-gel glasses can be produced with a wide variety of compositions and can be used to entrap a large number of different (bio)molecules, offering the possibility of tailoring the network structure, thickness, pore size and pore distribution by appropriate control of the variables involved in the preparation.

Finally, MIPs are one of the most efficient strategies that offer a synthetic route to artificial recognition systems by a template polymerization technique. MIPs are synthetic polymers containing imprinted nanocavities, which are able to specifically rebind their target. After careful optimization, their synthesis is usually comparatively straightforward and based on conventional polymer chemistry. A mixture of functional monomers and target molecules will form a pre-polymerization complex due to the non-covalent interactions between target and functional monomer. This complex is stabilized in an appropriate porogen. Crosslink monomers are added to this mixture to create a stable matrix. After radical polymerization, a bulk polymer is obtained containing the embedded targets. Subsequently, the target is extracted from the polymer matrix resulting in nanocavities, which act as the complementary binding sites. MIPs show similar specific binding characteristics and selectivity towards their target as antibodies and have the advantage of being robust and inert over a wide range of different environments. Furthermore, MIPs have the potential to be regenerated, which is usually not possible with biomolecules when using other types of solid supports. MIPs can be readily used for analytical separation, by packing them directly into suitable columns. A criticism of MIPs is the large quantity of template molecules required to prepare a useful amount of imprinted polymer, a potential problem if the target analyte is expensive or difficult to obtain; however, this problem is compensated by their high stability.

Different applications of MIPs for FTO have been proposed. Valero-Navarro et al. applied MIPs selective for monoamine naphthalenes (1-naphthylamine and 2-naphthylamine) for their respective or total determinations [14]. The synthesized polymer using naphthalene as a template molecule was ground and sieved to obtain the particles of 80 to 120 μm in diameter before use. Because of the similar π-π interactions or the formation of hydrogen bonding between MIPs and the analyte, the selectivity for 1-naphthylamine and 2-naphthylamine is not very high. The interference from 1-naphthalenemethylamine, which is also adsorbed on these MIPs, is serious and could not be eliminated. On the other hand, 1-naphthol and 2-naphthol, which are considered as most relevant potential interferents and actually adsorbed on MIPs, did not cause serious interference at the wavelength where 1-naphthylamine and 2-naphthylamine are measured. On the other hand, in the case of a water-soluble template molecule, a MIP for the determination of nafcillin in a milk-based product was also applied [15]. Although most of the α-lactamic- and anthracycline-based antibiotics did not interfere, only ampicillin caused an error when its molar ratio to nafcillin was more than 5.

2.4. MEASUREMENT OF THE ANALYTE AND REGENERATION OF THE SOLID SUPPORT

FTO can be based on (a) the measurement of an intrinsic analytical property of the analyte (e.g. absorbance, fluorescence) so avoiding any derivatization reaction, or (b) the measurement of an analytical property of the derivative product obtained after reaction between the analyte and a derivatizing reagent.

In the first case, although no derivatizing reagent is used to make the procedures more selective, performing the on-line spectroscopic measurements directly on the solid support increases the selectivity towards the targeted species compared with conventional flow analysis with the detection step performed in homogeneous solution. They are the simplest type of FTO, especially if they do not need solutions besides the carrier (no additional eluting solution is used, so providing longer sensor lifetime and baseline-signal stability).

The number of sensors based on monitoring a derivative product after derivatizing reaction is quite a lot lower than those measuring the native signal from the analyte(s). Two different options are possible in this configuration: (a) on-line derivatization, before the sample plug reaches the solid-phase area, and retention of the derivative species on the sensing zone (the most usual design in this case is one merging zone with double synchronized injections of both sample and reagent), or (b) immobilization of the reagent on the sensing support, generating the reaction product when the sample plug reaches the sensing microbeads, so successively reutilizing the immobilized reagent on the solid phase.

When a strong interaction between the analyte and the solid support is produced, it makes the elution of the analyte species from the solid phase very difficult. The implementation of a regeneration step is an important and key requirement in this case. It can be achieved by using an eluting solution, which could be an aqueous solution containing a high concentration of organic solvent, aqueous solutions of different pHs or even a pure solvent (alcohol, acetone, etc.). This eluting solution produces swelling and compactation of the resin beads and can generate small bubbles which will have a great effect on the solid phase light measurements. Nevertheless, in some cases complete elution cannot be done even if such kinds of desorbing agent solutions are used. Bead injection spectrometry (BIS) is very effective for this kind of cases. This concept is explained in next section.

Chapter 3

3. BEAD INJECTION SPECTROSCOPY

In this methodology the disposal of the solid support beads and the injection of new sorbent beads into the flow cell are carried out after each analytical cycle. The BIS concept, first proposed by Ruzicka et al., was introduced to use together with SIA [16]. It uses minute amounts of beads on which the reagent is adsorbed and the analyte is preconcentrated and monitored via optical fibers in specially designed (jet-ring-configured) flow cells. The bead suspension is pumped into the cell and beads are retained while solution goes through a circular gap narrower than the diameter of the beads. After measurement is performed, beads are automatically discarded.

The disposal of the adsorbent beads and the injection of new sorbent beads into the flow-through cell are carried out after each analytical cycle. A bead injection (BI) cycle with its different steps is shown in Figure 6. The name "jet ring" cell comes from the configuration and operation of the bead retention cell made as part of the flow conduit by inserting a solid rod to block the beads in the assigned space. The channel of the cell has an inner diameter slightly larger than the outer diameter of the blocking rod, forming the O-ring gap between the rod and the wall of the channel all around the rod. Only solutions are able to flow out while beads whose sizes are larger than the gap are packed due to the continuous incoming flow pushing the beads against the blocking rod. Used beads are discarded by the jet of solution reversely and rapidly flowing upstream.

Because the fresh adsorbent beads can always be used for both the retention and the optical sensing of analyte, this is applicable to target analytes which are strongly adsorbed on the solid phase and when the regeneration of the solid phase is difficult or the matrix species in samples are irreversibly adsorbed and lower the adsorption capacity of the adsorbent beads. Many kinds of adsorbent beads for the selective adsorption of target analytes such as Alizarin Red S or Zincon-loaded anion exchanger (Sephadex QAE A-25), protein G-coated Sepharose 4B beads and Cytodex microcarrier beads on which mouse embryonic fibroblasts are grown, have been applied to BIS.

Some advantages in comparison to conventional FTO derive from its renewable nature, and they include: (1) suitability for implementing reagent-based assays without requiring full reversibility of the sorption/elution process; (2) long-term operation because sorbent compaction and clogging do not take place; and, (3) preservation of sorption behaviour because surface contamination and deactivation do not occur. However, two fundamental prerequisites to be fulfilled for the bead material to be explored in a renewable fashion are the feasibility of forming a stable suspension and reproducible manipulation throughout the automated system. Consequently, it is highly recommended ensuring bead-size homogeneity and spherical shape of the reagent-supporting entities in order to prevent compact settlement in the channels of the assembly. Hence, conventional, reversed phase, chemically modified, silica-gel lumps are not really suitable for this purpose as a result of their irregular shape and size distribution.

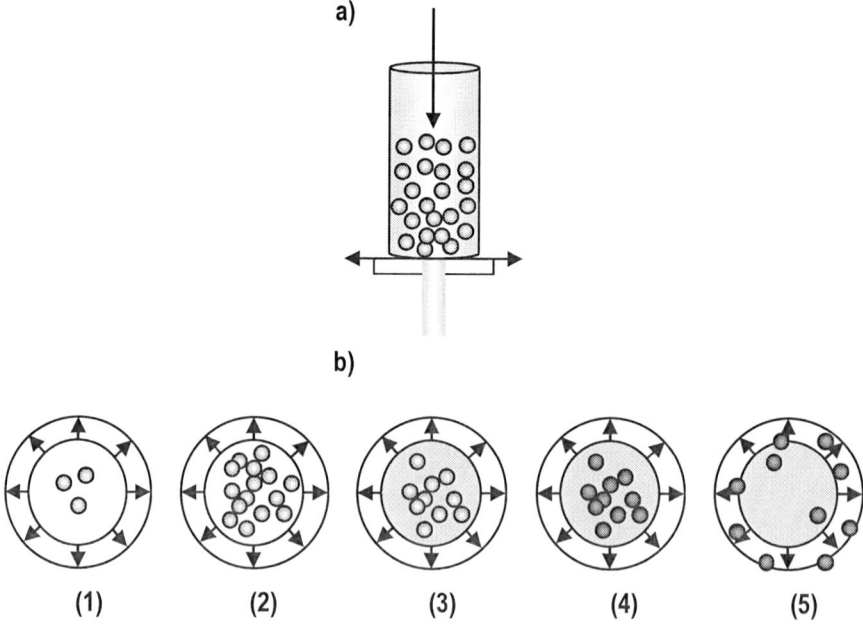

Figure 6. Jet ring cell diagram: (a) Front view; (b) Cross section (different steps)

In spite of the advantages of these systems, some laboratories may not afford this relatively high-cost instrumentation, while a simpler FIA system implemented with the use of commercial flow can be as appropriate as SIA methodology is for this BIS approach. This configuration has been developed by Ruedas Rama et. al., so obtaining different FTO. The procedure for each analysis (Figure 7) consists in the introduction of an exact volume of the bead suspension in the flow cell of a FIA system. Once the beads are trapped in the flow-cell and the baseline established, the sample is injected and the analytical signal obtained on the solid support. Finally, flow direction is reversed by means of a second peristaltic pump and the beads are automatically discarded and transported out of the system.

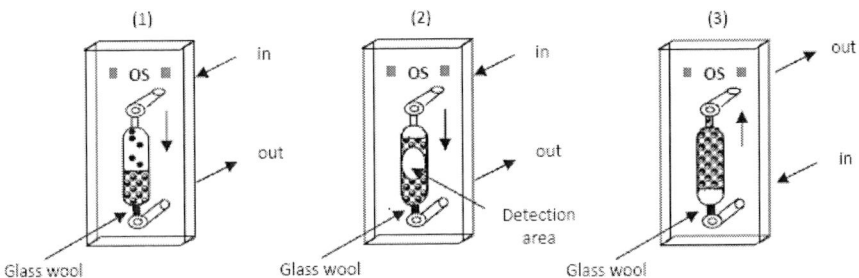

Figure 7. Different steps of the bead injection suspension into the flow cell: (1) Injection; (2) Detection; (3) Elimination.

To demonstrate the utility of this approach, the system [17] shown in Figure 8 has been applied to the measurement of the absorbance of the complex formed between Fe(II) and ferrozine (Fz) (sodium 3-(2-pyridyl)-5,6-diphenyl-1,2,4-triazine-4'-4''-disulphonate).

The procedure comprises five steps. (1) An exact volume of a bead suspension with the reagent immobilized on the surface is injected by means of the injection valve V1, loaded into the flow cell (see Figure 7, step 1). (2) The beads are trapped in the cell and perfused with the carrier stream, and the baseline for the subsequent absorbance measurements is established. (3) The sample (Fe(II)) is injected via valve V2 and the analyte reaches the bead surface; it is chelated and the signal is continuously monitored (see Figure 7, step 2). (4) The reaction between reagent and analyte on the beads is accomplished. (5) Finally, beads are automatically discarded from the flow cell at the end of the assay cycle by reversal flow (see Figure 7, step 3). All these steps can be seen in the corresponding fiagram (Figure 9) which is registered along all the process.

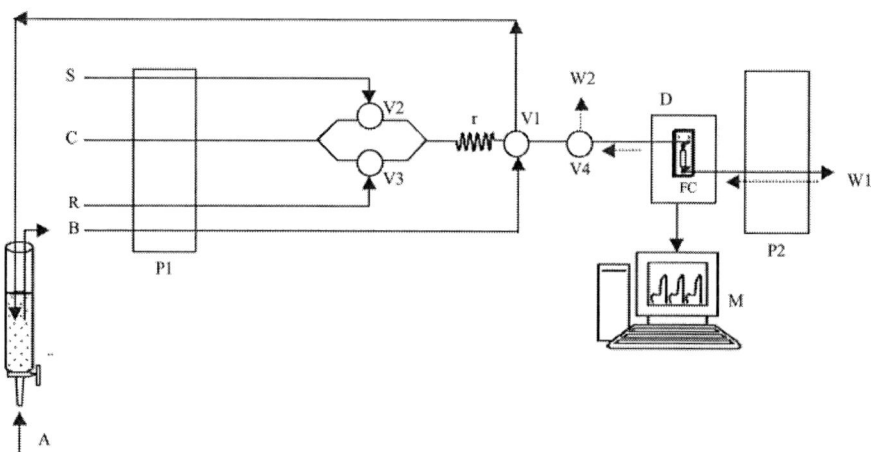

Figure 8. Manifold used in a BI system. A: air; S: sample; P1, P2: peristaltic pumps; C: carrier solution; R: reagent solution; B: bead suspension; V1, V2, V3: injection valves; V4: selection valve; r: reaction coil; D: detector; FC: flow cell; W1, W2: waste; M: computer.

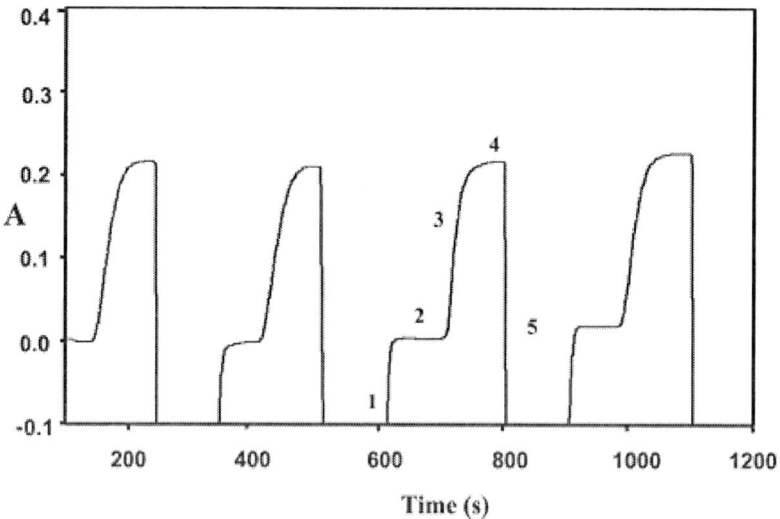

Figure 9. Fiagram obtained in a BI system.

A chromatography glass column is used in order to obtain a homogeneous aqueous suspension of the microbeads by purging air gently through it. This column is illustrated in Figure 10.

Magnetic beads could also be used in the development of renewable surface optosensors. This innovative application and manipulation will give new freedoms to SIA or FIA using this new material. Since magnetic beads can be trapped in a flow system using a magnet, their retention in the flow cell as well as their discharge can be easily accomplished with an (electro)magnet. Besides, magnetic beads are high-potential materials for a fully automated detection system because they have a large surface area and allow separation of target molecules from the reaction mixture.

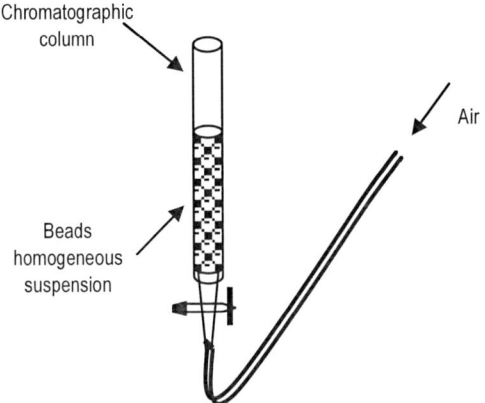

Figure 10. Schematic diagram of chromatographic column.

Finally, a further improvement in the automation of the microfluidic handling of suspended beads as renewable carriers of reactive groups or immobilized reagents was gained through the introduction of the novel Lab-on-Valve (LOV) principle [18]. It is based on integrating a set of microchannels within a miniaturized system placed atop a conventional multi-position valve, thus allowing many unit operations to be implemented readily. This rigid, compact SI-LOV coupling behaves as a portable, versatile laboratory, which is specially adapted for real-time sorbent-extraction optosensing schemes via in-valve integrated microcolumn reactors equipped with optical fiber-detection facilities.

Chapter 4

FLOW-THROUGH CELLS

Selecting a suitable flow-cell configuration for the implementation of FTO is not a trivial task. In order to achieve high concentration factors, the target compound should be sorbed preferentially on a small portion of the support material. In contrast to batch applications where a uniformly covered sorbent material is brought into the cell, in flow systems, the immobilized analyte forms a longitudinally distributed concentration gradient. The profile of the concentration gradient is governed by several factors, including the distribution ratio of the compound between aqueous and solid phase, the capacity of the sorbent material, the packing density and homogeneity of the particles, and the geometrical dimensions of the sorbent. Under favorable conditions, the target species is retained in a narrow zone at the flow-cell head.

With respect to the optical measurements, the ideal situation is when the area illuminated by the incident light beam matches the preconcentration section. As long as this is ensured, light attenuation for a given chemical assay and flow-cell geometry depends only on the amount of compound retained on the column. A mismatch will cause either deviation from the Lambert-Beer law and reduced dynamic range, when the illuminated area is smaller than the retention zone, or an increase in the background attenuance, when the illuminated area is larger. Besides, the transmitted light beam should be geometrically as similar as possible to the sorption region to avoid sensitivity decreasing through optical dilution of the analyte (i.e., diminution of the effective concentration on the solid-phase detected by the optosensor). The highest sensitivity for a particular mass loading would theoretically be achieved when the light is focused on an infinitely thin sorbent layer on which the sample flow impinges.

Direct proportionality between light absorbance and path length does not occur in flow applications, because a larger amount of solid material is required for a thicker layer, so the concentration factor is reduced. In addition, increasing the path length to 1-2 cm (without considering multi-reflection effects) increases light scattering, which, in turn, adversely affects the signal-to-noise ratio and the precision of the transmittance measurements. Furthermore, the performance of the FTO deteriorates through the build up of back-pressure. However, path lengths below 1 mm create a large longitudinal distribution of the analyte and cause low breakthrough volumes, so they are not too recommended. Two main aims must be pursued when a suitable cell for the design of a flow-through sensor is being selected: (1) the concentration of the monitored product on the support in an area as small as possible of it; and, (2) the incident light beam must be focused to this area without loss of light to the surrounding zone. The best results are provided by cells with short path lengths (1-1.5 mm), which ensure compatibility between the system and the detector and prevent the species from lying outside the irradiated area.

Different types of flow-through cells, originally designed for liquid-phase measurements, have been employed for solid-phase measurements. The most appropriate commercial flow-through cell for spectrophotometric measurements in UV region is the Hellma 138-QS cell (1 mm light path, 50 µl inner volume) (Figure 11). The mentioned flow-cell is also used with Fourier transform Raman detection.

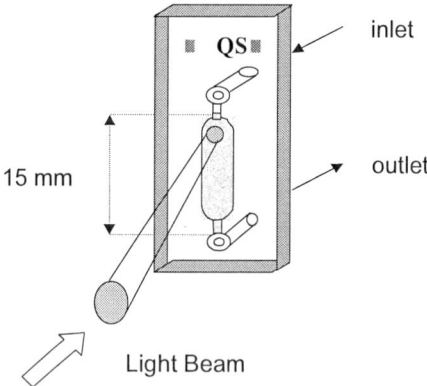

Figure 11. Flow-through cell (Hellma 138-QS) filled with support.

Higher levels of the packing material into the cell would imply that the support zone where the species of interest is sorbed would fall outside the detection area and so, a lower and wider signal would be obtained; with lower levels, the light beam would pass through the solution completely or partially and, consequently, a decrease in the signal would be obtained. Therefore, the top of the resin is kept as close as possible to the light beam, this latter being completely covered by the resin.

In the case of CL measurements, the Hellma 138-QS cell was the first commercial flow-cell employed in this kind of optosensors. However, the Hellma 137-QS (1 mm light path, 260 µL inner volume) has also been tested and a high increase in sensitivity was observed when compared to the first one. Normally, the solid support, inside the flow-cell, is situated in front of the window of the photomultiplier tube (PMT) to obtain the maximum signal (Figure 12).

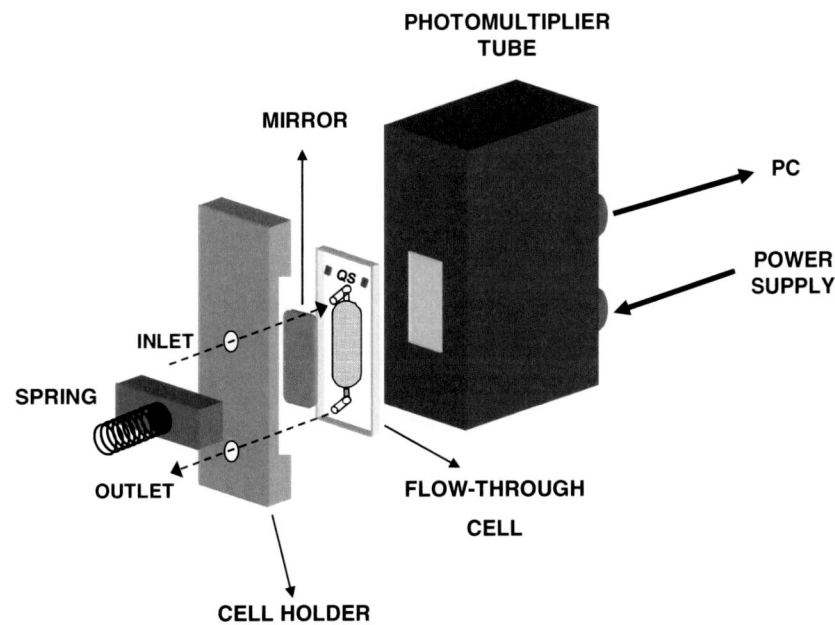

Figure 12. Components of a home-made FTO with CL detection.

For fluorescence, phosphorescence and lanthanide-sensitized luminescence (LSL) measurements, the Hellma Model 176.052-QS flow-through cell (25 µl inner volume) with a light path of 1.5 mm is usually employed (Figure 13). In this case, in order to secure that the species of interest retained by the packing solid material is in the light path, the solid support level is maintained just some millimeters beyond the cell window. It is necessary to take into account that, when an eluting solution other that carrier one is used to desorbs the species of interest from the solid support, ion-exchange resins suffer alternately swelling and compaction, so altering the level of the support in the cell and the baseline. This is due to the different chemical nature and concentration of the carrier and the eluting solutions. Hence, in this case, and for preventing compactions lowering the support level below the light beam, the flow cell must be filled up passing the eluting solution through it. In every case, it is necessary to condition the solid support by passing the carrier solution through it for a few minutes.

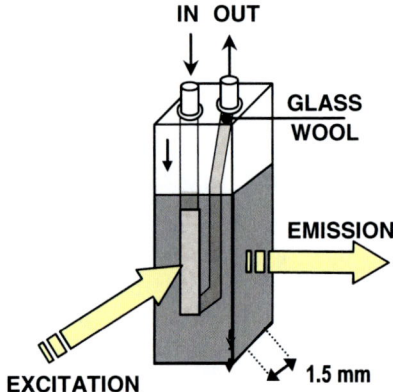

Figure 13. Flow-through cell (Hellma 176.052 QS) filled with support.

All these cells are blocked in the outlet with glass wool to prevent particle displacement by the carrier stream while the inlet is kept free. Obviously, in every case the optimum level of solid support in the cell depends on the shape and height of the light beam and so, it depends of the instrument used for measurements. It is worth mentioning that a simple glass column or methacrylate modules in different shapes can also be used to retain the solid microbeads.

Chapter 5

5. FLOW METHODOLOGIES IN FTO

In recent years more and more strict regulations related to the quality control of different compounds (drugs, pesticides, metals, etc.) led to increasing demands on the automation of analytical assays carried out in appropriate control laboratories. The main aims of recent developments in flow analysis have been to progressively achieve lower reagents consumption, higher repeatability, complete automation of the analytical procedure and miniaturization.

Different requirements need be attained for direct solid-phase measurements in flow systems:

- ✓ *Thermodynamic of the retention process.* The coefficient of distribution of the retained species needs to be considered, i.e., analyte retention on the solid support needs to be favorable. With this aim, non-polar materials are used for the retention of hydrophobic species, and so on. When reaction occurs at the solid surface (e.g., complex formation with an immobilized ligand), the equilibrium constant and the reaction conditions (e.g., pH and presence of masking agents) should also be taken into account. However, in flow based systems the residence time is usually lower than those required to achieve the steady state, and the thermodynamic values are only indicative.
- ✓ *Kinetic aspects.* Analyte retention needs to be compatible with the short residence time characteristic of the flow methodologies. Fast retention of the analyte in comparison to that of interfering species can also be exploited to improve selectivity.
- ✓ *Stability of the solid support.* The solid-phase should not change significantly when submitted to the different media required in the

retention and elution steps. This is more critical when a reagent is immobilized on the solid support and leaching can reduce the lifetime of the solid-phase sensor. This is often minimized by proper immobilization of the reagent at the solid support and selection of suitable carrier and eluting solutions. The stability of the solid-phase under irradiation should also be evaluated, especially for measurements based on fluorescence, when the solid surface is submitted to high power radiation favoring photo-degradation.

- ✓ *Reversibility*. the solid-phase needs to be efficiently regenerated by the eluting solution after sample measurement by the eluting solution, making feasible the use of the same solid support for several measurements. One alternative is replacement of the whole solid support in each measurement cycle, which can be also efficiently carried out in flow systems.
- ✓ *Compatibility between the solid support and the measurement system.* Materials which cause excessive attenuation of the radiation beam (by absorption or scattering) are not suitable for spectrophotometric measurements based on transmission, but can be used for measurements by reflectance or fluorescence.
- ✓ *Backpressure*. The measurement cell, particle size, and packing need to be carefully selected to avoid fluid leakage. This aspect generally limits the total flow-rate that can be employed, also affecting system design. As an alternative, reagents can be immobilized in porous membranes, minimizing the drawbacks related to the increase of the hydrodynamic impedance.
- ✓ *Selectivity*. The solid-support and adequate conditions should allow separation of the analyte from interfering species. This is often achieved by chemical derivatization with a selective reagent immobilized at the solid support or previously to the retention of the species at the solid surface. Control of the composition of the medium (e.g., pH) and use of selective eluting solutions are other commonly employed approaches.

Main basic components of any flow system include propelling system, reaction manifold, flow cell (seen in previous Section), detector and sample introduction device.

Propelling system. Pumping systems can be categorized as piston driven, peristaltic, or gas driven. The gas driven pumps have been used in process monitors. The advantages are that the system is pulseless, inexpensive, and that there are no moving parts in the pump, therefore less down time. The disadvantages are that variable flow rates will be created by the changing resistance in the reagent(s) and carrier storage bottles. Moreover stopped-flow type measurements are difficult to perform. Piston driven pumps are reasonably reliable and can handle nonaqueous solutions. The disadvantages are that they are expensive compared to the other pumping systems, pulsations created by these pumps must be dampened, stop-flow is unlike, and the number of moving parts is significant. Peristaltic pumps appear to be the most useful for flow methods (Figure 14). They are reliable, programmable, have few moving parts, and are relatively pulseless. The disadvantages are that the pumps are moderately expensive and, more important, the pump tube must be replaced routinely. Anyway, peristaltic pumps are used in most of the systems developed up to date.

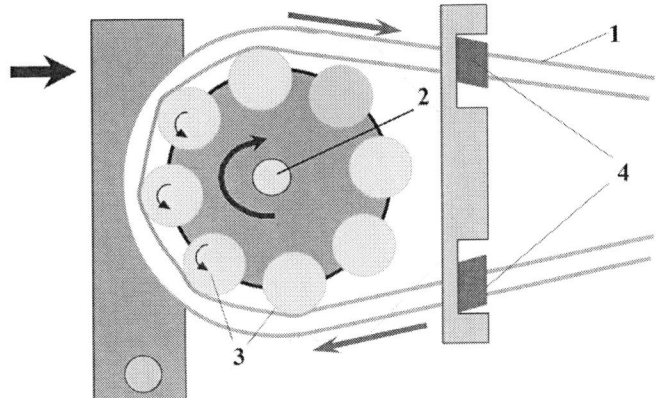

Figure 14. Scheme of a peristaltic pump. (1) thermoplastic flexible tube; (2) rotor; 3 (rollers); (4) stoppers. The arrows indicate the sense of liquid circulation.

Reaction manifold. The tubing materials which have been used up to date are teflon and stainless steel. The last one is used primarily because of its ruggedness. The negative point about stainless is that the analyst cannot see through the tubing as is in teflon. Teflon tubing is rigid and reasonably rugged. The chemistry is easily observed since the teflon is transparent enough to see through. The negative to teflon can be situations where there is a large amount of protein in the stream. Teflon will absorb protein to its surface. If the surface coat is undesirable, the teflon is clearly a bad choice for tubing material.

Another tubing parameter to be considered is the inside diameter. Traditionally the tubing i.d. has been 0.5 mm, although 0.8 mm is also widely used.

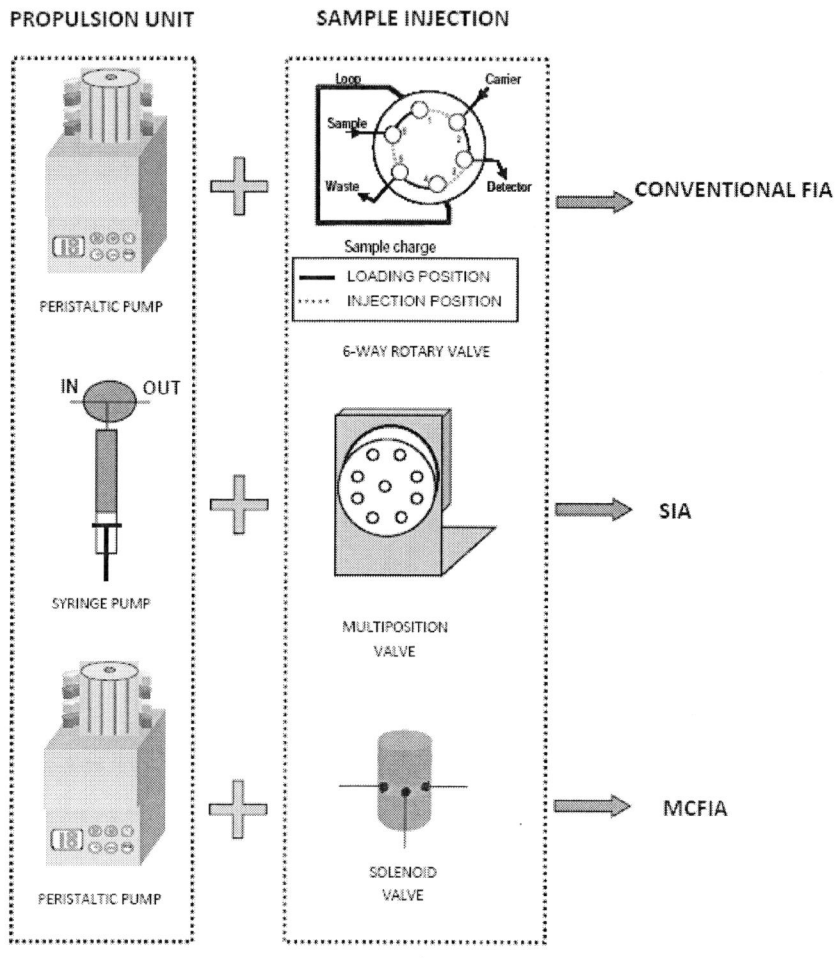

Figure 15. Scheme of flow methodologies employed in FTO.

Detector. Essentially any existing flow through detector can be used with a flow system. Spectrophotometry, nephelometry, fluorescence, chemiluminescence, atomic absorption, flame photometry, potentiometry with ion-selective/modified electrodes or field effect transistors, amperometry with sensors and biosensors and voltammetry with wire-type or rotating disk electrodes are the most important detection techniques used in flow methods. Even some detectors such as mass spectrometry have been employed. The types of detection more widely used in FTO will be studied later.

Sample introduction device. As it was indicated in the Introduction, the most important flow methodologies are FIA, SIA and Multicommuted flow techniques, each method presenting advantages and drawbacks when compared to each other. They use different sample introduction devices (Figure 15) although the main aim, in all of them, is to replace operations usually carried out by the analyst, improving precision and reducing the analysis time.

5.1. FLOW INJECTION ANALYSIS

FIA is based on the insertion/injection of a liquid sample into a moving non-segmented carrier stream of a suitable liquid. The injected sample forms a zone, which is transported by the carrier through a coil of tubing to a detector. The detector measures a physical parameter of the sample (absorbance, fluorescence, etc.) that changes continuously as a function of time as the sample passes through the flow cell. This means that the concentration of the species being monitored is continuously changing with time. The carrier may contain a reagent that reacts with the analyte to yield a detectable product, or may consist of an inert solution and in this case the carrier serves as a means of transporting the sample to the detector. Thus, the FIA response curve is a result of two processes, both of kinetic nature, the *physical* process of dispersion of the sample zone within the carrier stream, and the *chemical* process of the formation of chemical species.

In the FIA technique, the basic components are a peristaltic pump to propel the sample and reagents, a series of plastic tubes to carry the solutions, injection valves to introduce constant volumes of sample and reagents in the system, and the detector. A diagram with these components is shown in Figure 16. The main advantages of a FIA system are repeatability, rapidity and versatility.

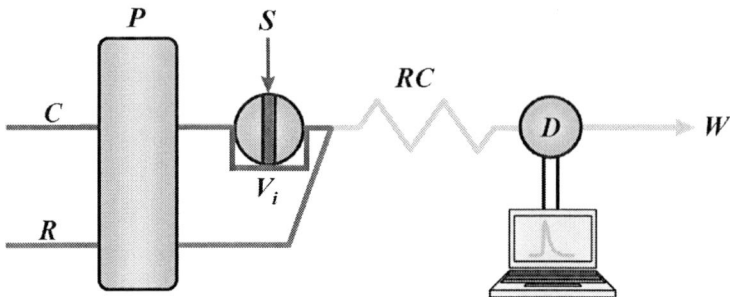

Figure 16. Diagram of a FIA system. P: pump; V: injection valve; RC: reactor (reaction coil); D: detector; C: carrier; S: sample; R: reagent; W: waste.

A schematic representation of the process which happens in a FIA system is presented in Figure 17. The analytical signal can be explained in three steps: (a) establishment of baseline and insertion of the sample aliquot; (b) retention of the analyte at the solid support yielding the analytical signal; and (c) elution process and regeneration of the solid support. In the scheme, a transient signal is obtained because the eluent reaches the solid support immediately after the whole sample zone cross the flow cell.

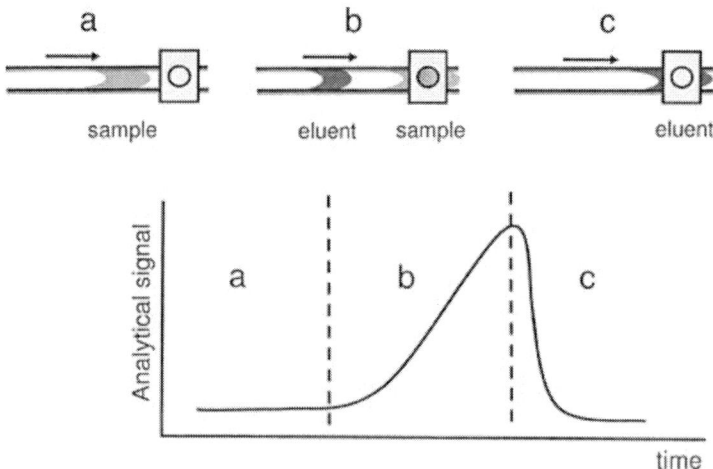

Figure 17. Representation of the analytical signal in a typical FIA-SPS procedure.

The injection valve mode approximates the plug injection and is a more facile way of inserting well-reproducible sample volumes into the carrier without disturbing its motion. Figure 18 shows a rotary injection valve and its operating mode. The loading and injection steps employed by displacing a movable part between two resting positions are a common feature of these devices. Air bubbles and pressure surges must be avoided during the injection because they will modify the pattern of the flow in FIA system, affecting dispersion and precision.

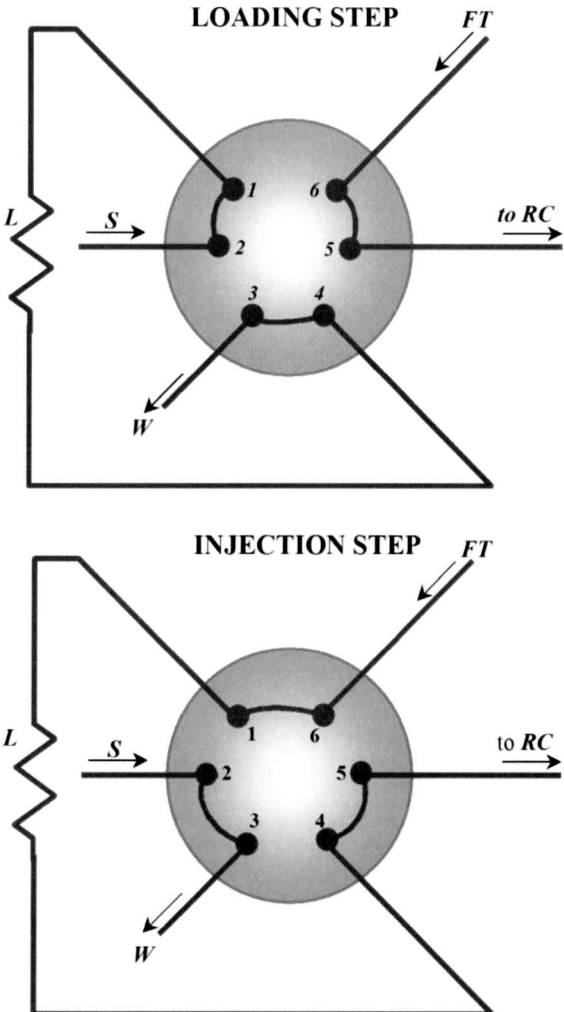

Figure 18. Six-way rotary valve injection procedure. S: sample; R: reagent; FT: carrier; RC: reaction coil; L: valve loop; W: waste.

This valve can be also used as selection mode. In this way, the configuration of the valve is totally different (Figure 19).

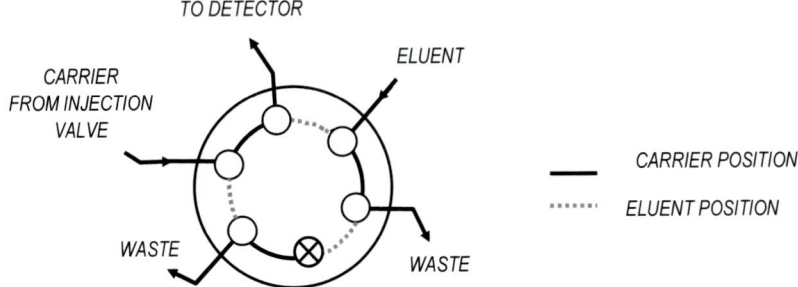

Figure 19. Six-way rotary valve selection procedure.

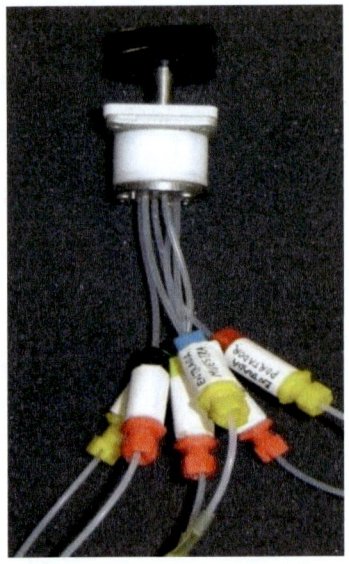

Figure 20. Six-way rotary valve.

The valves usually employed in FIA systems are the manually-controlled six-way rotary valves (Figure 20). The sample charge and injection is performed by the movement of a mobile element between two fixed positions, being the sample volume selected by the length of the sample loop and the internal volume of the injection valve.

The tube placed between the injection valve and the detector represents the dispersion coil or the reaction coil and it has a uniform internal diameter (most frequently used being 0.5 or 0.8 mm). It usually is a Teflon tube tightly wound in the form of a coil to promote mixing. Reactors filled with chemically inactive glass beads have been used to improve the mixing without increasing the dispersion. In these rectors a well-controlled and reproducible dispersion pattern of the sample zone and high sample throughput can be achieved.

Next figures show some of the most usual FIA manifolds used in FTO (C: carrier solution; E: eluting solution; C/E: carrier/eluting solution; P: peristaltic pump; S: sample; IV: injection valve; SV: selection valve; D: detector; SZ: sensing zone; W: waste; r: reactor; R: reagent stream):

a) The simplest is a monochannel manifold in which the carrier solution acts also as eluting solution: as the sample plug tail reaches the sensing zone, the analyte is eluted from it, so originating a transient signal (Figure 21).

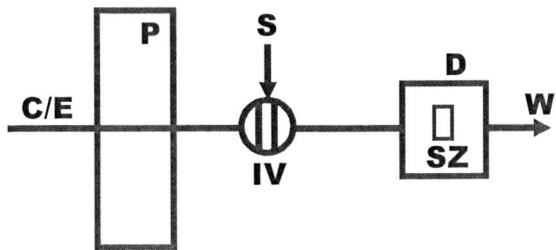

Figure 21. The simplest monochannel manifold.

b) The use of two different appropriate alternative carrier/eluting solutions by means of a selecting valve makes possible to perform the sequential determination of two analytes with a slight modification in the single channel manifold above described (Figure 22).

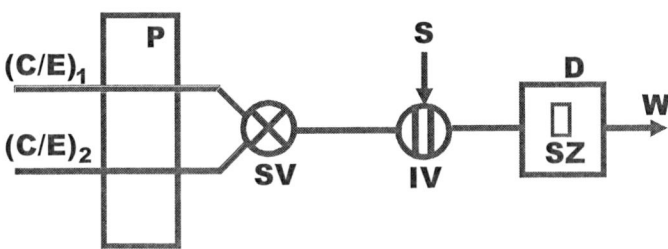

Figure 22. Manifold with two C/E solutions.

c) When the species is retained on the solid support so strongly than the carrier cannot elute it, an additional eluting solution has to be used. An additional injection or selection valve is then used (usually placed as near to the sensing (micro)zone as possible) to allow the eluting solution to act (two alternatives in Figure 23). This manifold usually shortens the lifetime of the sensor when an ion exchanger is used as support.

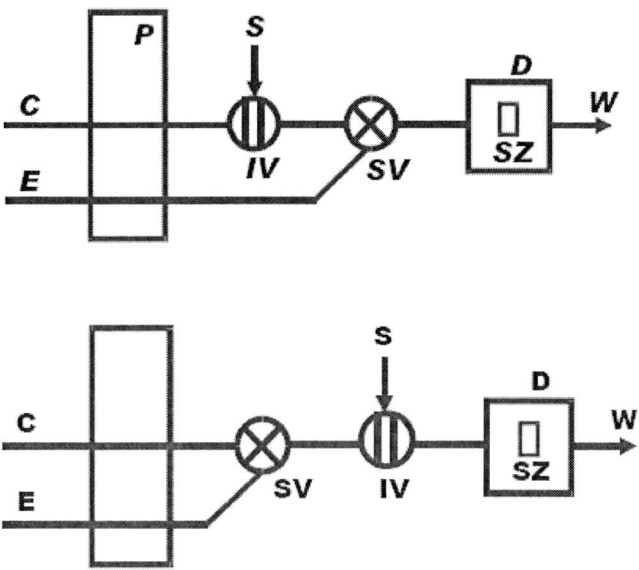

Figure 23. Manifolds with additional eluting solution.

d) A double synchronized injection of both sample and reagent is performed in those sensors in which a previous derivation of the analyte(s) is performed due to the analytical signal does not correspond to an intrinsic property of the analyte(s). A reaction coil is then used to allow the reaction takes place in it before the sample plug reaches the cell (Figure 24). A modification of this manifold has also been used in order to perform the determination of several analytes. The use of a previous selecting valve before injecting the sample in the carrier stream (Figure 25) allows to alternatively select the reagent to be injected in a synchronized way with the sample.

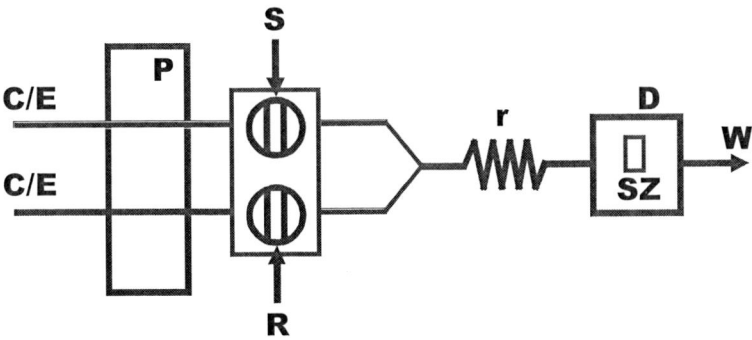

Figure 24. Manifold with previous derivation of the analyte (I).

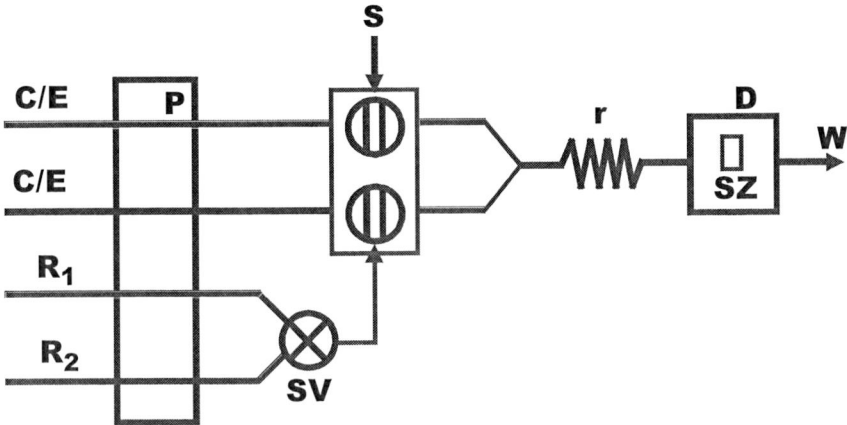

Figure 25. Manifold with previous derivation of the analyte (II).

e) Manifold shown in Figure 26 has been used for the direct determination of an analyte or the indirect determination of another one. By actuating the selection valves SV_1 and SV_2 it is possible either to insert directly the sample stream into the carrier solution by means of the injection valve (direct determination of the first analyte) or previously to merge with a reagent stream (to favour the hydrolysis of the second analyte, so rendering the first one) along the reactor, which is immersed in a thermostatic bath (B) (indirect determination).

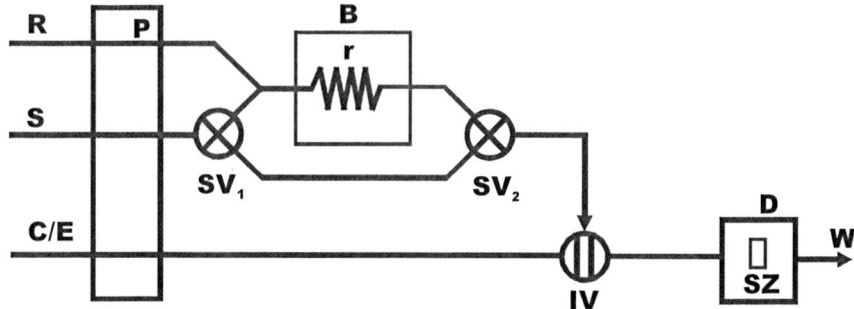

Figure 26. Manifold for direct or indirect determination of the analyte.

f) Finally, in manifold depicted in Figure 27 the loops L_1 and L_2 are filled with eluting and sample solutions, respectively. By sliding the commutator central bar, sample and eluent aliquots are simultaneously inserted into the carrier solution. The reactor provides a proper time delay to avoid excessive carryover between sample and eluent zones. So, the elution is always performed by using just a definite volume of regenerating solution.

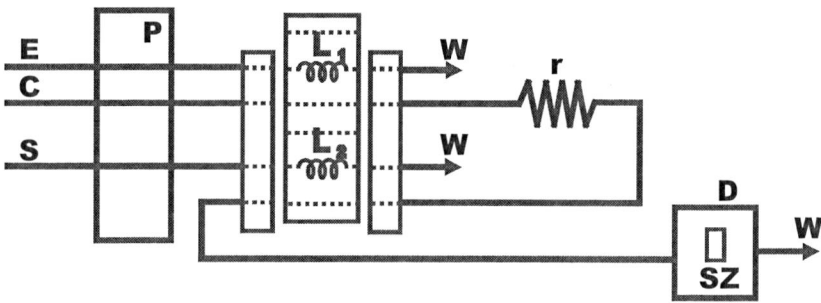

Figure 27. Manifold with commutator central bar.

5.2. SEQUENTIAL INJECTION ANALYSIS

SIA was introduced as a following generation in the development of flow injection technique. This is based on the same principles as FIA, namely controlled partial dispersion and reproducible sample handling, and it offers different possibilities with a series of advantages and disadvantages in relation with its parent technique. The instrumental simplicity, robustness, ease and efficiency with which hydrodynamic variables can be controlled, and high flexibility and modes maintenance requirements of this modified technique have turned it into a very popular choice with both research and industrial analysis laboratories. On the other hand, SIA has proven to be a technique that can be designed to operate in a multi-parametric way, which is of special interest when considering the design of the environmental monitors. However, in spite of these advantages, SIA presents two major disadvantages: the sample throughput is lower than that of the usual flow systems and major difficulties in the mixture of sample and reagents.

A SIA system is assembled using a multi-port (usually 10 ports) electrically activated selection valve, a high-precision peristaltic or syringe pump, a suitable flow-cell detector, tubing/reaction coils and connectors (as those used in FIA) and a personal computer. A diagram is shown in Figure 28.

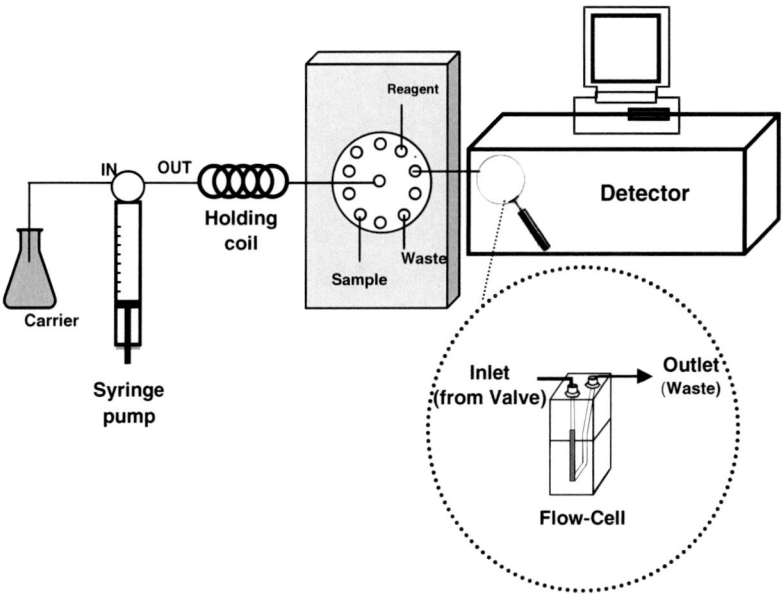

Figure 28. Diagram of a SIA system.

Appropriate software must be available to control the flow direction, rate and timing of the pump, the position of the multi-port valve and to collect and process the data. The heart of the system is a multi-port selection valve, used in place of a conventional injection valve and this is the primary difference between FIA and SIA. This valve (Figure 29) delivers accurately measured volumes of carrier, sample, standards and reagent solution to a holding coil by connecting its common port to a reversible pump featuring a precisely controlled forward-stop-backward motion. The common port can access any of the other ports by electrical rotation of the valve. The holding coil placed between the valve and the pump prevents the aspirated (injected) solutions from entering the pump.

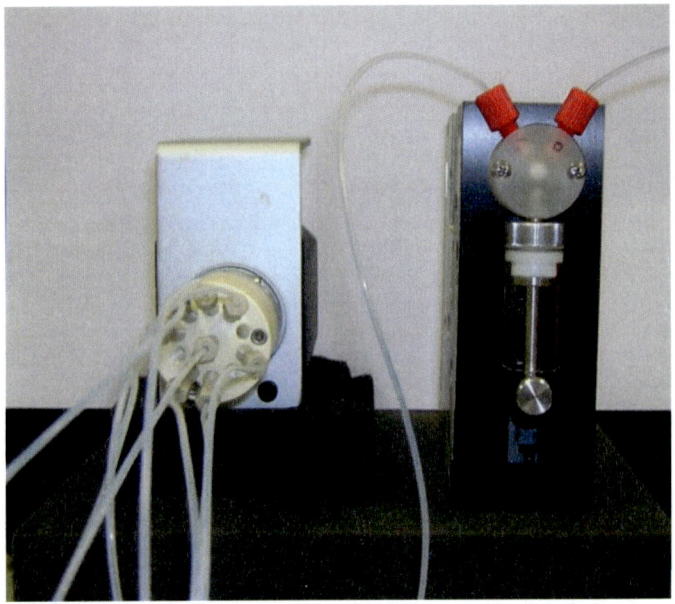

Figure 29. Multi-port selection valve and syringe pump.

An operational manifold design for SIA is illustrated in Figure 30. Initially, the system is filled with washing or carrier solution, which is aspirated into the holding coil by the pump moving in a forward motion. Each measuring cycle begins by switching the multi-port valve to the sample line and aspirating a precise measured volume (few μL) of sample into the holding coil by the pump moving in a backward motion; the pump is stopped during the rotation of the valve to avoid pressure surges. Next, the valve is switched to the reagent line and a precisely measured volume of reagent is drawn into the holding coil. Thus, the sample and reagent solutions are sequentially injected into the holding coil next to one another, hence the name of this technique. A second reagent may be aspirated on the other side of the sample.

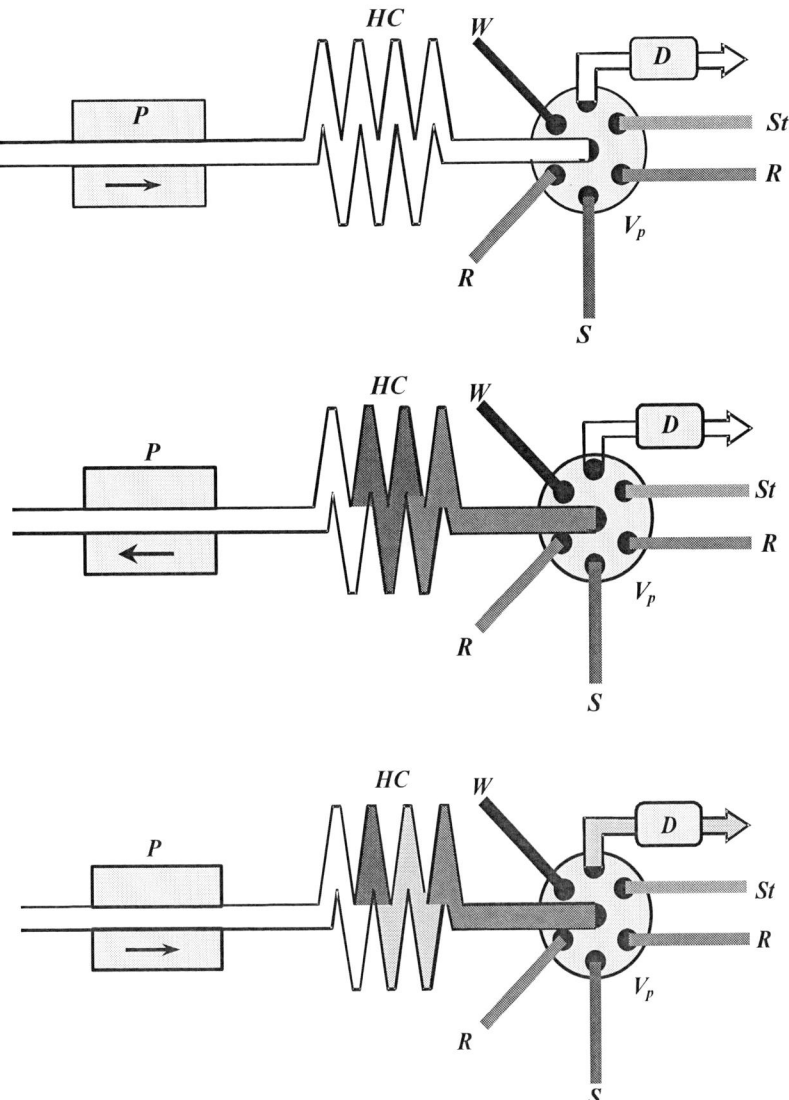

Figure 30. Diagram of a SIA system. P: single-line reversible (high-precision bi-directional) pump; HC: holding coil; V_p: multi-port valve; D: detector; S: sample; R: reagent; St: standard solution; W: waste.

Finally, the valve is switched to the detector port and the pump propels the sequenced zones forward through the reaction coil to the detector. The cores of sequenced zones penetrate each other via laminar flow and diffusion. If the radial mixing is promoted by a suitable choice of coil geometry, the analyte and reagent zones mix and produce detectable species and a transient signal as in conventional FIA is recorded.

The complex reagent/product/sample zone can be either transported through the detector continuously, or stopped within detector, resulting in measurement of the rate of formation of the reaction product. In this way, kinetic information can be extracted from the SIA signal. There is no limit to how many solutions or devices (reaction coils, mixing chambers and detectors) can be nested around the multi-port valve. A series of standards can be permanently nested around valve, being ready for automated recalibration whenever is necessary.

The biggest advantage of SIA over FIA is that is not necessary the physical reconfiguration of the flow path. The injected sample volume, reaction time, sample dilution, reagent/analyte ratio or system calibration are controlled from a computer keyboard. Indeed SIA is fully computer-compatible and allows the configuring of the system to perform complex chemistries. Moreover, other advantages of SIA are very simple manifolds (all solutions handling is achieved by means of the multi-position valve), minimum solutions wasting (the piston pump only works during the time strictly needed to aspirate the amount of sample and/or reagent needed), high repeatability, possibility of using very low volumes of solutions and very robust systems. Although implementation of SIA with SPS is a recent and promising methodology in FTO, a small number of SIA optosensors have been developed.

5.3. MULTICOMMUTATION

Multicommutation consists of the employment of discrete commutation devices (solenoid valves, for example) to build up dynamic manifolds that can be easily reconfigured by software. This approach greatly increases the versatility of the flow systems since each analytical step can be independently implemented. The main difference between multi-commutation and the other flow methodologies is that insertion volumes are replaced with insertion times. Among the multicommuted flow techniques, Multicommuted Flow Analysis (MCFIA), Multisyringe Flow Analysis (MSFIA), and Multipumping Flow Systems (MPFS) are the most widely employed nowadays. They combine the advantages given by FIA such a high injection throughput with those given by SIA such as a minimal reagent consumption and waste generation.

MCFIA systems are typically constituted by a peristaltic pump and a set of three-way solenoid valves, automatically controlled by appropriate software, which can be arranged creating a flow network. Each valve can adopt two positions, being the whole system assimilated to an electronic circuit with a variable number of active nodes. MSFIA relies on a device designated by a multisyringe burette which is a multiple channel piston pump, driven by a single motor of a usual automatic burette and controlled by computer software through a serial port. And finally, MPFS systems are based on the use of a series of individual solenoid micro-pumps acting as liquid propelling units. These solenoid micro-pumps are responsible for sample-reagent introduction and manifold commutation. All these methodologies will be explained in more detail.

5.3.1. MCFIA

MCFIA technique was devised by B.F. Reis et al. in 1994 in the Piracicaba CENA (Brazil), being the key element in these systems the solenoid valve, usually 3-way solenoid valves. These valves act as a switch between two different positions, "OFF" and "ON", remaining permanently connected two of the three valve ports. The switching procedure of the three-way solenoid valve is shown in Figure 31. There are two main configurations for the design of a MCFIA system: aspiration of the solutions or propulsion of the solutions. When the solutions are aspirated (Figure 31(A)), one valve can be used for controlling the flow of two different solutions. As it can be observed, one port is always opened, and it can be communicated with one of the other two ports by means of the up and down movement of the solenoid; therefore, the solution required is selected by activating or deactivating the 3-way solenoid valve.

On the other hand, when the liquids are propelled, only one solution is controlled by each individual solenoid valve. In Figure 31(B), the scheme for the carrier solution flow is shown. When the valve is deactivated, the carrier flows towards the detector; however, when the solenoid valve receives the electric pulse, the solenoid moves up and the carrier is recycled to its corresponding vessel. In the case of sample and reagent solutions, the recycling and detector ports would be exchanged. Hence, the sample/reagent solution would be inserted in the system only when the valve is activated.

Next, a representative MCFIA manifold employed for the determination of a target compound is depicted in Figure 32. V_i represents the 3-way solenoid valves; the straight line is the flow of solutions where the valve is OFF, while the dotted line is the direction followed by the corresponding solution when the valve is switched on.

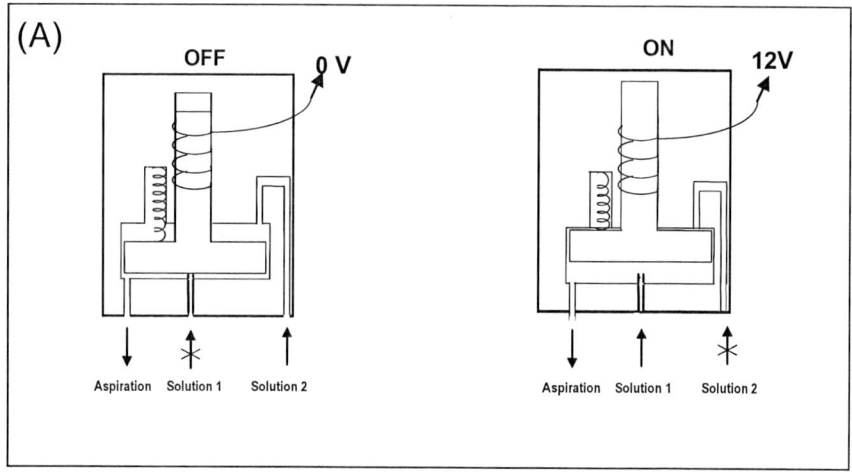

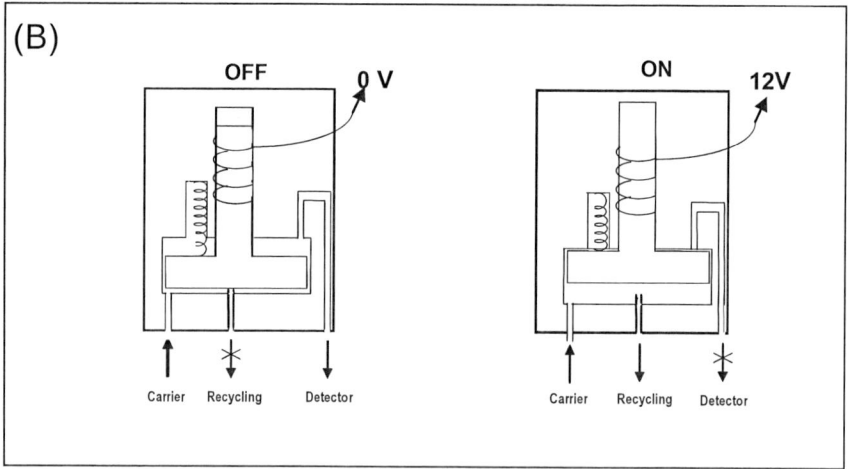

Figure 31. Procedure of the three-way solenoid valve.

The sample volume inserted is directly related with the time of the electric pulse, and it can be changed by modifying the insertion sequence profile. It is possible to insert the sample volume in one or several aliquots, therefore mixing the sample with the carrier solution or with reagents solutions. The volume of sample spent is only that one aspirated during the time the valve remains "ON". As a result, a high saving in terms of consumption of sample and reagents solutions is obtained by employing this automatic methodology. The early MCFIA systems used a single-channel propulsion system to aspirate the liquids via individual valves. However, when the liquids are aspirated into the system, the devices used tend to insert air bubbles in the system; as a result, liquid propulsion has been employed more recently instead of aspiration. Peristaltic pumps, automatic burettes or syringe pumps have been employed for propelling the liquids in these systems.

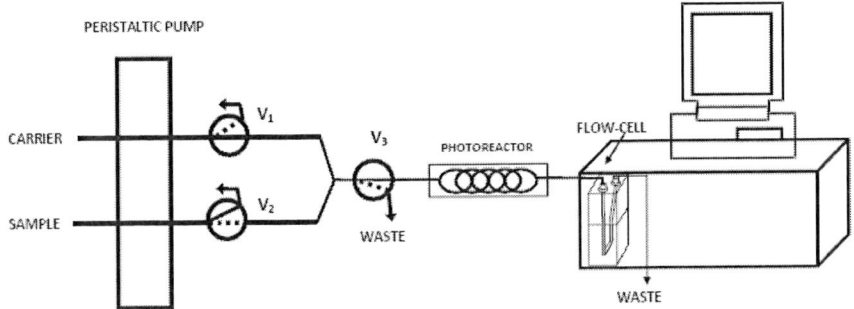

Figure 32. MCFIA manifold.

The use of flow networks comprising several solenoid valves acting as independent commutators (automatically controlled by appropriate software) allows the modification of the flow network without physically altering the valves position and/or connections, therefore widening the scope of flow analysis possibilities. Another advantage of these flow systems is the possibility of performing on-line dilution. A minor drawback of the solenoid valves is the damaging effect on the teflon inner membranes caused by the heat released by the solenoid coil when the valves are activated (ON) for a long time. This overheating can be avoided using electronic protection systems.

Although the vast majority of MCFIA manifolds comprise 3-way solenoid valves, 2-way solenoid valves have been also used. These valves are available in two different configurations: a) normally closed, where the flow of liquid is permitted only when the valve receives the electric signal; b) normally open, where the flow of liquid is allowed until the valve is activated and closed.

5.3.2. MPFS

These systems, developed in 2002 by two research groups at the Pharmacy Faculty of the University of Porto (Portugal) and the Piracicaba CENA (Brazil), are based on the combined use of solenoid valves and pumps, leading to the so called solenoid piston pumps or micro-pumps. The MPFS manifold is made up of a series of solenoid micro-pumps, as many as the solutions required by the analytical method. The micro-pumps operate in the form of pulses in a monodirectional way. Each pulse produces a piston stroke of the pump, which propels a preset volume of liquid (depending on the model of the micro-pump). The number of pulses defines the total injected volume and the flow-rate is determined by the stroke frequency. As a result, these systems are similar to a MCFIA system, where the peristaltic pump is replaced by the solenoid pumps, which also act as the solenoid valves themselves (opening or closing the flow of liquids). Due to the operation in pulse mode, the overheating prevention systems described for the solenoid valves is not required.

The principal advantages of a MPFS is its high flexibility, easy configuration, robustness and low cost (as explained, the solenoid pump operates as both the valve and liquid propeller). In addition, considering that each micro-pump acts individually in the propulsion of the fluids, distinct sampling techniques (such as the single sample volume, binary sampling or merging zones) can be explored in a versatile and independent way. The better mixing of reagents and sample due to the pulse mode provides higher peaks when compared with other flow techniques.

A typical and simple manifold is depicted in Figure 33, where only 3 solenoid pumps are used, which were responsible for the individually handling of three different solutions: carrier, reagent and sample. The baseline is established by inserting the carrier solution actuating P_1, while the reaction is produced by the actuation of P_2 and P_3. Tthe pumps provide the function of propeller system and solenoid valves at the same time.

As it has been previously stated, the combination of different flow techniques have been used if required for a specific application. The most common approach is using micropumps together with solenoid valves; normally 3-way valves are used, but pinch solenoid valves can be employed too [19]. A pinch-type valve consists of a solenoid magnetic actuator with a pinch plunger and a silicone tube; the movement of the plunger opens or closes the silicone tube, therefore controlling the flow of the solutions.

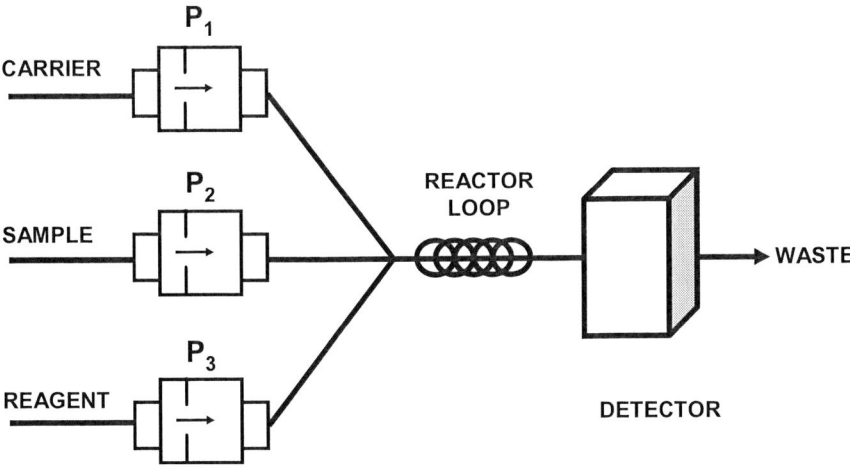

Figure 33. MPFS manifold.

5.3.3. MSFIA

MSFIA was developed by Cerdá et al. (Department of Analytical Chemistry, University of Balears Islands, Spain) together with the firm Crison (Alella, Barcelona, Spain) with the aim of combining the advantages of both SIA and MCFIA.

The device consists of a conventional automatic titration burette adapted in a way that the motor can simultaneously move the pistons of four syringes, avoiding the need of using separate burettes in parallel. This movement is accomplished by using a horizontal metal bar, moved by the motor of the burette, which accommodates the four syringes. In addition, each syringe has a three-way solenoid valve placed at its head, which is used to select the direction of the liquid flow. As the burette motor moves the four syringes at the same time, different flow-rates (values lower than 0.1 up to 72 ml min^{-1}) can be achieved in each channel by modifying the dimensions of the syringes. The multisyringe burette, which can be coupled with other methodologies, is shown in Figure 34.

The main advantages o MSFIA are: robustness of the system, due to the absence of pumping tubes (allowing the use of aggressive reagents and solvents), possibility of performing direct sample and reagent confluence (in classical SIA there is only one syringe) and its capacity to work under moderate backpressure. The propulsion system used opens up new possibilities, combining the multichannel operation of peristaltic pumps with the constant, pulseless and exactly known volume delivery achieved by piston pumps. However, MSFIA has the same disadvantage as for piston pumps that the forward movement must be stopped to reload the syringes, decreasing the sampling frequency. The coupling of MSFIA with MPFS has been introduced to avoid this problem.

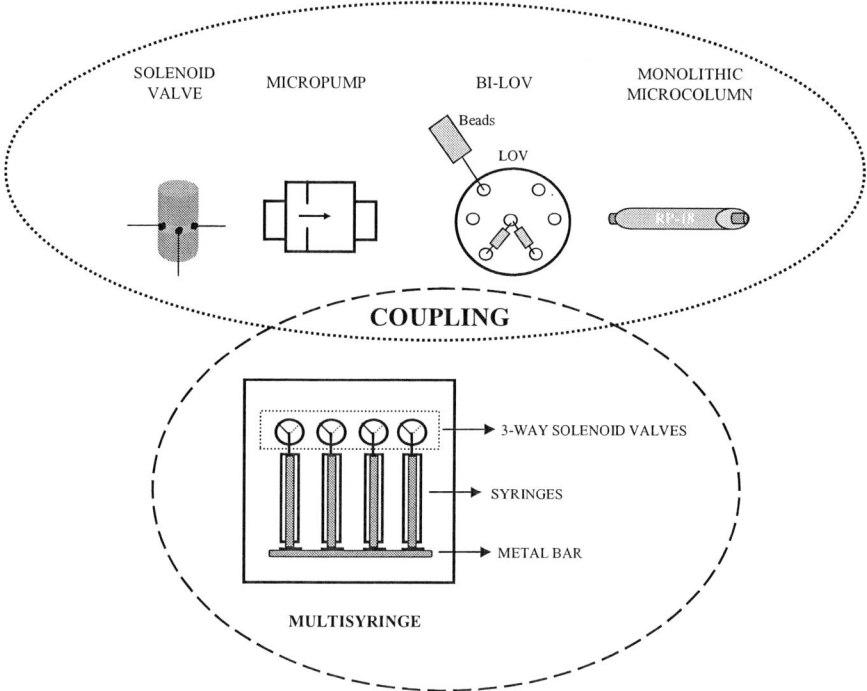

Figure 34. Coupling of multisyringe burette with other methodologies.

5.4. COMPARISON OF FLOW METHODOLOGIES

All the flow methodologies employed in the development of FTO have favorable intrinsic characteristics, as previously stated. However, each one presents specific advantages when compared to each other. A comparison between these methods in term of the degree of automation, repeatability, sensitivity and sample frequency is made. The introduction of multicommutation and SIA has provided additional automation when compared to FIA, due to the complete absence of human intervention during each analysis. In FIA, an operator has to switch manually the rotary valves in order to inject the sample or reagents solutions. In recent MCFIA and SIA optosensors, the flow of solutions is automatically controlled by appropriate software and the consumption of sample and reagents is lower, making these methods more suitable for routine analysis.

The repeatability of SIA and MCFIA methods is theoretically higher when compared to FIA ones due to the higher degree of automation. However, a comparison of the relative standard deviation (RSD) (%) reported in different published papers show similar results in all cases. This is due to the use of the solid support, which compensates the theoretically small differenced that could be found when working in homogeneous solutions. On the other hand, the sensitivity of MCFIA and conventional FIA methods is slightly better when compared to SIA. This is due to the dilution of the sample in the SIA system when all solutions have to be aspirated previously to be pumped towards the detection area. For the same reasons of the aspiration/pumping process in SIA, the sample frequency in these methods is lower when compared to MCFIA or classical FIA.

In general, the incorporation of multicommutation and SIA to flow-through optosensing has provided the laboratories with favorable tools in order to develop automatic, rapid and environmental-friendly methods for routine analysis.

Chapter 6

CLASSIFICATION OF FTO IN TERMS OF NUMBER OF ANALYTES

Most of FTO developed are mono-parameter or mono-sensing optosensors, i.e. the sensing zone responds only to one analyte in the sample. Nevertheless, the novel introduction of optosensors responding to more than one analyte from the same sample, which are called multi-parameter or multi-sensing optosensors, expanded their range of applications. Most of these systems have dealt with the determination of two analytes; however, some systems have been designed in order to quantify three analytes too.

6.1. MONO-SENSING OPTOSENSORS

In this type of flow-based optosensors, the solid support responds only to an analyte in the sample. The determination of this latter can be carried out with or without the previous on-line generation of a derivative product. The direct measurement of a property of the analyte itself is the simplest approach used in the development of flow-based optosensors and it constitutes an interesting contribution to green analytical chemistry since it does not need solutions besides the carrier and (sometimes) eluting solutions.

The first photometric sensor of this group described for the determination of an anionic active principle was for ascorbic acid [20]. An acetate buffer solution is used as carrier/self-eluting solution and an anion exchanger gel as sensing solid support. Three different calibration ranges are obtained from three different sample injection volumes. The determination of paracetamol [21] is also based in the same principle but using NaCl/NaOH (pH=11.0) as carrier/self-eluting solution, because the analyte needs a higher pH value than ascorbic acid to get deprotonation and retention on an anion exchanger gel.

Fluorescence is the most frequently used detection technique in this type of mono-sensing optosensors. Nevertheless, as native fluorescence is not a common characteristic of organic compounds, different strategies have been used for the determination of non-fluorescent molecules, such as chemical derivation or UV-irradiation. Three fluorometric mono-parameter sensors have been developed related to vitamins B: pyridoxine [22], riboflavin [23], and pyridoxal [24]. Two of the former are based in the measurement of the native fluorescence of the analyte (pyridoxine and riboflavin) and another one (pyridoxal) uses a derivative reagent. Pyridoxine was determined by transitory retention on Sephadex SP C-25 using NaCl/HCl as carrier/self-eluting agent. Riboflavin was determined by measuring its native fluorescence at 515 nm retained on a hydrophobic solid support (C_{18} silica gel) and transported by a phosphate buffer (pH = 6.0) solution. And pyridoxal was determined via formation of a fluorescent complex with beryllium in ammonia buffer and measurement of the fluorescence of the derivative product at 450 nm.

6.2. MULTI-SENSING OPTOSENSORS

These systems, which respond to more than one analyte, are more difficult to develop, as more complex series of requirements have to be accomplished for this purpose. Four different strategies have been used in the development of this kind of flow-based optosensors, which have allowed the simultaneous determination of two or three analytes.

6.2.1. Use of two different sensing zones

Two options are available in this case: the employment of two different detectors or the use of a double beam spectrophotometer. In the first case, the determination of vitamins C, B_2 and B_6 was carried out. Taking into account the native fluorescence of B_2 and B_6, they were determined by means of a spectrofluorometer. However, as vitamin C does not display native fluorescence, its reaction with permanganate in a sulphuric acid medium was performed and the obtained CL was measured with a luminometer. The sample plug was easily directed to the required detector by means of a SIA system [25].

In the second case, using of two different sensing (micro)zones and a double beam spectrophotometer, an injection valve is operated twice successively to inject the sample plug which is carried out alternatively each time to the appropriate sensing (micro)zone just actuating on the selecting valve SV_2 (Figure 35). The appropriate carrier stream is selected each time by means of the selecting valve SV_1. Thiamine and ascorbic acid were quantified in pharmaceutical preparations following this procedure. The vitamins were concentrated on ion-exchange gels, thiamine on Sephadex SP C-25, and ascorbic acid on Sephadex QAE A-25. Both solid supports were packed in two different flow cells and the absorbance of the analytes measured by means of a double-beam spectrophotometer [26].

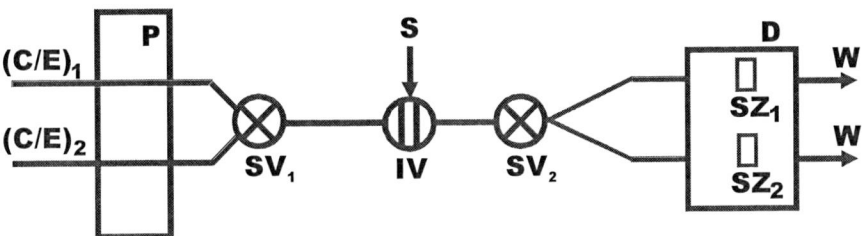

Figure. 35. Manifold using two different sensing (micro)zones.

6.2.2. Separation in a minicolumn

Usually, the separation of the analytes is performed using a mini-column filled with the same solid support used in the flow-through cell and placed just before the cell. These minicolumns are home-made, in glass, and they are placed just before the flow-cell. The length of the material packed is a key variable since it has to be sufficient as to allow a complete separation of the analytes in the minimum possible time. Appropriate working conditions have to be established as to allow the selective retention of one (bi-parameter sensor) or two (three-parameter sensor) of the analytes on the solid support packed into the minicolumn, while the other one(s) reach(es) the detection area. This requires a carrier/eluting solution allowing different retention-elution kinetics of the analytes in the minicolumn. After the measurement of the first analyte, the use of additional eluting solutions allows desorption of the analyte(s) retained in the minicolumn and their transport to the sensing zone.

This approach (see Figure 36) was used for the determination of binary and ternary pesticide mixtures in commercial preparations and water samples [27].

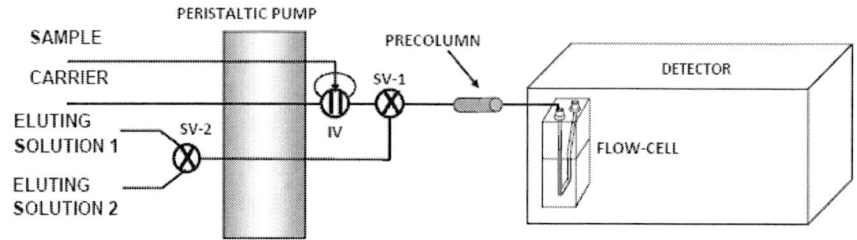

Figure 36. Manifold using separation in a minicolumn.

Thiabendazole and metsulfuron methyl were simultaneously determined in water samples by placing in the flow system a minicolumn packed with C_{18} silica gel, the same solid support used as sensing zone [28]. Firstly, the sample was injected into a 15% (v/v) methanol solution, which originated a weak retention of thiabendazole in the minicolumn and its elution by the carrier solution itself, being monitored by measuring its native fluorescence. After this, metsulfuron methyl was eluted from the minicolumn with a 60% methanol solution and monitored after its UV-irradiation and generation of a strongly fluorescent photoproduct.

6.2.3. Separation in the flow-cell

A very simple design for the on-line separation of analytes, without involving the use of additional devices in the manifold, is the integration of the minicolumn into the same flow-cell. The minicolumn is replaced by the introduction in the flow-cell of an additional amount of solid support and, therefore, the level of support in that is higher than the usual one. The separation of the analytes takes place in the zone of the support above the detection area and allows a sequential arrival of them to this latter. This strategy involves some advantages like: (a) simplicity in manifold and procedure, (b) higher sensitivity, and (c) higher throughput.

Fuberidazole and o-phenylphenol could be simultaneously determined, despite their severe spectral overlapping, by using an additional amount of solid support (C_{18}) in the flow-cell and two different carrier/eluting solutions

(Figure 37). The use of a 30% (v/v) methanol solution originated a very strong retention of o-phenylphenol on the upper part of the support, in a zone situated above the irradiated one, while fuberidazole was transitorily retained and monitored. Then, a 60% (v/v) methanol solution allowed the elution of o-phenylphenol and its monitoring [29].

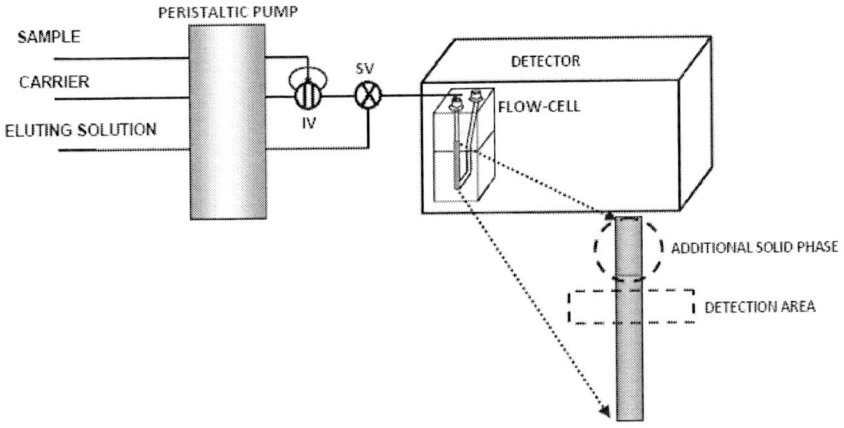

Figure 37. Manifold with additional amount of solid support in the flow cell.

Another interesting strategy in biparameter sensors with no derivative reagents are those based on the two following aspects [30]: (1) the measurement of one of the two analytes (which is not retained on the solid support) when it passes through the interstices among the sensing microbeads, and (2) the transitory measurement of the other analyte on the solid support as it passes thorough the sensing zone, being eluted by the carrier itself. These biparameter sensors are used when the concentration of the first analyte is much higher than the second one. Therefore, it is a sensor with a dual functioning of the sensing area: (1) detection in homogeneous solution with a low effective pathlength (first analyte) and (2) retention/elution/detection on the solid microbeads. This sensor could be used to determine solely each analyte in different separate samples (no minicolumn needed).

6.2.4. Separation with mathematical treatment

The simultaneous determination of two or three analytes can be also carried out without their previous on-line separation, but making use of a mathematical treatment of their analytical signals. Several chemometrics algorithms such as first-order multivariate calibration partial least-squares (PLS) algorithm and second-order algorithms such as multiway PLS (N-PLS) and unfolded PLS (U-PLS) have been applied for this purpose. These approaches can resolve overlapping signals and reduce interference problems as well as background noise [31].

The quantitation of ternary mixtures of analytes in pharmaceutical preparations, such as the mixtures of caffeine-acetylsalicylic acid-paracetamol [32] and caffeine-dimenhydrinate-acetaminophen [33] was achieved. The three compounds were simultaneously retained on the solid support microbeads (C_{18} silica gel) and their full UV spectra were recorded by means of a diode array spectrophotometer during the retention time. The extensively overlapped spectra of the analytes could be resolved by PLS regression. After collecting the response of the multisensor, its active microzone was regenerated by using methanol as eluting agent, leaving it ready for the next determination.

Chapter 7

DETECTION TECHNIQUES

7.1. UV-VISIBLE SPECTROSCOPY

In measuring the light attenuation by a solid phase, the biggest problem is that a relatively small net absorbance, caused by the colored species adsorbed on the solid phase, has to be measured under a fairly large background attenuance of the solid phase. This large background attenuance, which is characteristic of the light measurements of a solid phase, is due to a surface reflection or a diffuse reflection from the solid layer. For example, if the attenuance of a 1-cm thick cross-linked polystyrene-type ion exchanger packed into a quartz cell is measured against air as the reference, the light attenuance is about 3 in absorbance units even at 700 nm, at which there is no light absorption by the ion-exchange resin. This means that the very low absorption of light passing through the resin phase, which is above or near the limit of detection of a common spectrophotometer for quantitative light measurements, has to be measured. In the case of batch SPS method, the light path length of a solid particle layer in the cell is 1 to 2 mm. Inevitably 50 to 1000 mL sample solutions had to be used for trace analysis to compensate for the shorter light path of solid particle layers.

In the case of FI-SPS using a micro flow-through cell, light attenuation by the cell also accompanies the background light attenuation of the solid phase. The use of respective optical fibers for incident light radiation and for collection of the light transmitted through the solid phase has attracted increasing interest for this purpose. If the cross-sectional area of the luminous flux of the fibers for incident light beam radiation is smaller than that of the flow-through cell, the light attenuance by the cell can be negligible. When the optical fiber is set as close to the solid particle layer as possible, a considerable amount of the light scattered by the solid particle layer can also be recovered. In focusing the transmitted light on a photomultiplier or on a grating for photo diode array detection, the selection of the proper accumulation time for the transducer is very effective for the reduction of background light attenuation by the solid phase.

UV-Visible spectroscopy has been traditionally the most frequently used detection technique in analytical laboratories due to its high flexibility for adaptation to a wide variety of analytical problems and the low-cost of equipments. However, not many analytical methods rely on the intrinsic absorbance of the analytes, as their intrinsic light absorption occurs at the UV region, where other co-existing compounds also absorb, interfering in the analytical measurements. As a result, UV region doesnot usually present the required selectivity for conventional spectrophotometric analysis. However, the implementation of the active solid support in spectrophotometric flow systems strongly enhances selectivity, excluding from the detection area all those species that cannot be retained in the experimental conditions. Different analytes have been determined simply by measuring their native absorbance. However, derivatization has been frequently required, especially when determining inorganic species, such as Zn in pharmaceutical preparations [34] or Cu in urine samples [35].

7.2. LUMINESCENCE

The use of luminescence detection techniques has gained considerable ground in analytical chemistry due to the high sensitivity and selectivity that can be achieved. Moreover, the combination of the FI method with luminescence detection is becoming increasingly important in clinical, biological and environmental analyses because of its wide linear dynamic range, reproducibility, simplicity, rapidity and feasibility.

However, not all luminescence techniques are selective enough, and different strategies can be employed in order to overcome some limitations, such as spectral overlapping, scattered light, quenching processes or background signal. The techniques employed in FTO have been fluorescence, phosphorescence, CL, and LSL, being the first one the most used up to date. All of them will be commented below.

7.2.1. Fluorescence

In the case of fluorescent light measurement emitted from the solid phase in the cell, utilizing a longer light path cell is not effective because the deeper part of the solid particles in the cell cannot be irradiated by the excitation light beam. To increase the excitation light beam intensity is also unsuitable for sensitivity enhancement because the scattered or reflected excitation light enters the emission light detector even if it is located perpendicular to the excitation light beam, which results in a serious emission background error. However, because the sensitivity of fluorometry is inherently higher than that of absorptiometry, a solid phase light path of 1 to 1.5 mm is sufficient for trace analysis, and the measurement of emitted light from the solid phase can be easily accomplished in a way similar to that with solution spectrometry. About 60% of the recent papers on FTO are concerned with fluorometry probably because of not only the simplicity of these methods but also the incompatibility of the sensitivity with the troublesome light measurements of solid phase absorptiometry.

Compared with solid phase absorptiometry, the applicability of solid phase spectrofluorometry, especially for the metal ion determination, has been limited because the number of fluorogenic reagents selective for a specific metal ion was small. However, many researchers have tried to develop this method for analysis of organic compounds such as pesticides, carcinogenic polycyclic aromatic hydrocarbons, pharmaceuticals, etc.

The added economy and simplicity of fluorometry and its relative flexibility have turned it into the officially recommended choice for determining the purity of many drugs in pharmaceutical preparations. The intrinsic fluorescence of the analyte has been used for the determination of several compounds, such as naphazoline [36] using a commercial solid support, tetracyclines using a modified cyclodextrin based sensor [37] and digoxin by means of a MIP-fluorosensor [38].

As there is a small number of compounds that exhibit native fluorescence, derivatization reactions have been employed to increase the number of applications. One novel approach is the so called photochemically-induced fluorescence (PIF). It consists in the generation of fluorophores from non-fluorescent analytes by on-line UV irradiation and presents inherent advantages over ordinary chemical reactions such as quicker reaction rate, fewer chemicals involved and smaller dilution factor. The photodegradation can be carried out by inserting a photoreactor just before the detector. This latter is constructed by coiling a PTFE tubing (0.8 mm i.d.) around a low-pressure mercury lamp (8-15 W, 254 nm), which is placed into an aluminium box to permit the maximum reflectance of UV light and heat dissipation. An important aspect to take into account in this type of optosensors is that the regenerating solution (carrier itself or eluting solution) has to be able to elute not only the species monitored but other products of the UV irradiation of the analyte, which possibly can remain retained on the solid support and interfere the next determination.

The implementation of PIF in FTO was described for the determination of thiamine in pharmaceuticals and human biological fluids [39]. In a recent paper, binary mixtures of sulfonamides were determined in urine samples making use of both native fluorescence and PIF [40]. The resolution was accomplished by means of a mini-column placed in the flow system; one of the sulfonamides was not retained in the mini-column and determined by measuring its native fluorescence on the sensing microbeads, whereas the other one was retained and, after its elution, photochemically converted into a strongly fluorescent photoproduct, which was transitorily retained on the sensing support in the flow cell and its signal monitored.

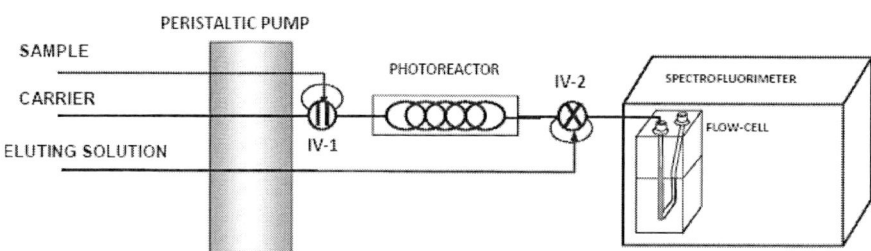

Figure 38. Manifold with PIF detection.

The typical flow configuration used when PIF is employed is depicted in Figure. 38. This configuration has been applied to the determination of the pesticide imidacloprid in peppers [41]. The irradiation of the sample solution was performed in the photoreactor, which was prepared by coiling PTFE tubing around the low-pressure mercury lamp. In this case, an additional eluting solution was required in order to regenerate the solid support microbeads after the signal was recorded.

7.2.2. Phosphorescence

When phosphorescence is selected as detection technique, a common aspect is the need for some form of molecular immobilization and/or protection in order to minimize non-radiative decay of luminophores, collisions with solvent, or the possibility of photochemical reaction.

Under a specific condition phosphorescence from a chemical component can be observed even at room temperature. This long-lived emission is called as room temperature phosphorescence (RTP). When comparing the RTP method with fluorometry, the RTP method has some advantages over fluorometry: the short-lived background luminescence or scattered light can be easily discriminated because of the long-lived emission of the analyte species, and very good separation of the maximum wavelength of phosphorescence from that of the excitation spectra due to Stoke's shifts is achieved.

The optosensing system for RTP from the solid phase is not very different from that for solid phase spectrofluorometry except for choosing the respective proper gate and delay times for the light detector to isolate the RTP signal.

RTP in fluid solutions can be observed in the presence of micelles, heavy atoms, oxygen scavengers such as sodium sulfite and nitrogen, etc. They play an important role in facilitating the energy transition from excitation singlet state to excitation triplet state of the phosphorescent species. In addition to micelles, cyclodextrins, vesicles or microemulsions have been used as ordered mediums to minimize self-quenching and to organize reactants on a molecular level, therefore increasing the proximity of heavy atoms and analytes. The RTP based on the use of a heavy atom is called heavy atom induced-RTP (HAI-RTP). In spite of the advantages of the RTP method over fluorometry, not many reports on the application of RTP have been published [42]. The main reason is the lack of an RTP system selective for a specific chemical component.

7.2.3. Chemiluminiscence

CL is a phenomenon involving the emission of light (usually in the visible or infrared (IR) region) as a result of a chemical reaction. Essentially, the process by which the luminescence is produced is identical to that for fluorescence except that no excitation light source is necessary. The analytical interest of CL arises from the ability to produce fluorescent molecules with no prior irradiation, thereby avoiding various problems derived from light scatter, unselective excitation or light source instability.

Until several years ago, CL flow sensors were relatively rare as compared to other type of optical flow sensors and most of them were limited to biosensors based on immobilized oxidases with dissolved CL reagents which react with hydrogen peroxide released from the enzymatic reactions to produce a CL light signal. In recent years, CL flow systems with immobilized or solid state reagents have received much attention and many analytical applications have appeared in the literature [43].

Recent developments of CL sensors are reviewed in various publications [44]. However, a serious limitation is the requirement to continuously deliver the CL reagent into the reaction zone because the CL reagent is consumed during the CL reaction, which is undesirable not only for the simplification of the detection device, but also for the cost, environment and resources considerations. An effective approach to solve this problem is to employ CL reagents in immobilized or solid-state format. In these systems, analytes are detected by the CL reactions either with the immobilized reagents directly or with the dissolved reagents which are released from the immobilized substrates. The use of sol-gel to immobilize enzymes has become a recognized process for preparing CL sensors. The key advantages of sol-gel are that there is little or no structural alteration of the encapsulate species and it is suitable for optical sensors due to its optical transparency and chemical stability.

Some basic requirements must be met by a substrate with immobilized reagent for application in a CL sensor system: (1) the substrate should have high mechanical stability and do not swell in aqueous solutions; (2) there would be low band broadening coupled with minimum back-pressure in a packed bed reactor; (3) for a long lifetime, a substrate with a high surface reagent coverage is preferable; (4) when the immobilized substrate is packed in a flow cell placed in front of the detector, a favourable transparency is essential for high sensitivity; (5) the substrate should exhibit good chemical stability, even at elevated pH conditions, which are frequently required to enable the light-producing reactions.

The development of immobilization techniques has provided the introduction of enzyme reactors which can be positioned before the CL reaction takes place, thus avoiding the lack of selectivity that may occur when a given CL reagent yields emission for a variety of compounds. In this alternative procedure, the analyte is the substrate of the enzymatic reaction and one of the products will sensitively participate in the CL reaction. Substrates that have been detected in this way include glucose, cholesterol, choline, uric acid, amino acids, aldehydes and lactate which generate H_2O_2 when flowing through a selective column reactor with immobilized oxidase enzymes in the presence of the necessary oxidant, usually O_2, present in the samples. Luminol, in the presence of a peroxidase catalyst seems to be one of the best systems for this post-column H_2O_2 determination.

A FTO sensor has been described for the determination of analgin based in the CL reaction of this compound and dissolved oxygen sensitized by Rhodamine 6G in the presence of acidic Tween 80. Rodamine 6G was immobilized on a cation-exchanger column placed in front of the detector [45].

Molecular imprinting technique is a rapidly developing technique for the preparation of polymers that would be used as sensing materials to design CL sensors. The imprinted cavities of a defined shape and functional groups in the MIP are expected to develop not only with the molecule recognition function but also as a special CL reaction medium. In a recent publication, an amoxicillin-imprinted polymer was synthesized with methacrylic acid as functional monomer and ethylene glycol dimethacrylate as cross-linker for the determination of the analyte in urine samples [46].

7.2.4. Lanthanide-sensitized luminescence

The trivalent cations of the lanthanides have photoluminescent properties that are favorable for several kinds of applications. However, it is difficult to generate this luminescence by direct excitation of the lanthanide ion, because of the ions' poor ability to absorb light. Nevertheless, they possess unusual spectroscopic properties when chelated with appropriate organic ligands. These chelates display a well-defined luminescence characterized by narrow and highly structured emission bands, large difference between absorption and emission wavelengths (Stokes's shifts) and long excited-state lifetimes. As a result, these chelates have been widely used with chemical and biological applications, mainly for the determination of organic analytes. This detection technique is called LSL.

As a general rule in LSL, once the chelate is formed, the lanthanide ion luminescence originates from an intramolecular energy transfer through the excited state of the ligand (organic analyte) to the emitting level of the ion. This luminescence process is not diffusion-controlled because the ion combines with the organic ligands by coordinate linkage. Tb (III) and Eu (III) are generally preferred and used in practical applications because of their longer decay times and intense luminescence intensities. Two main approaches are used in order to increase the efficiency of the energy-transfer process: a) use of organized media, such as micelles and cyclodextrins; and b) use of a solid support to retain the organic ligand (analyte) and the lanthanide ion. In both cases, the sensitivity and selectivity of the system are increased.

The energy-transfer process can be intramolecular or intermolecular.

Intramolecular energy transfer. This process occurs when the lanthanide ion forms a chelate with an organic ligand. The intramolecular energy transfer from the triplet state of the organic ligand to the lanthanide ion, depends on the structure of the ligand and the position of its triplet state, and could lead to a marked increase in the luminescence intensity. This phenomenon is called the "antenna effect" and such complexes are considered to be light conversion molecular devices because they are able to transform light absorbed by the ligand into light emitted by the ions. The excitation-emission mechanism is shown in Figure 39.

The organic ligand (the analyte) absorbs energy, leading to an excited singlet state (S_1), which then goes over to an excited triplet state (T_1) and the energy is intramolecularly transferred from this triplet state of the ligand to the localized intra-4f shell energy levels of the lanthanide ion, which is excited and emits its characteristic radiation. As a result, the excitation and emission wavelengths of the system are specific of the ligand and lanthanide ion respectively.

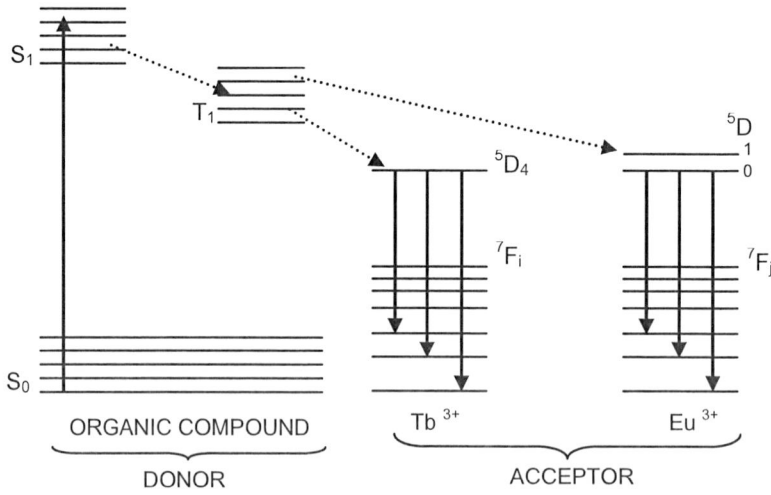

Figure 39. Mechanism in intramolecular energy transfer.

Intermolecular energy transfer. Population of luminescent levels of rare earth ions is also possible through energy transfer from excited, non-chelated organic molecules. An extensive investigation of energy transfer from excited aromatic aldehydes and ketones to lanthanides has shown that the donor is the triplet state of the organic compound. In homogeneous fluid solutions and in the absence of an acceptor, the triplet state is deactivated without phosphorescence since the non-radiative deactivation rate is usually much higher than the rate of phosphorescence. In the presence of a suitable acceptor, energy transfer can successfully compete with the radiationless decay processes of the analyte. The transfer process is diffusion controlled and energy is transferred after an encounter between acceptor and donor species, i.e., as a collisional interaction. Energy transfer from molecules in the lowest triplet state is favored not only by the longer natural lifetime of this state, but also by the spin invariance of the transfer mechanism from the sensitizer to the solvated lanthanide ion, which changes its spin quantum number by unity on excitation from ground state ^7F to an excited state ^5D. Therefore, the process appears to be analogous to the intermolecular transfer of energy between organic triplets in solution.

The use of a solid phase to retain the lanthanide-analyte complex offers the advantage of avoiding the use of these enhancers or surfactants, therefore minimizing costs and making the methodology environmentally-friendly. A particular group of compounds that have been determined by TSL optosensing are the broad-spectrum antibiotics quinolones. Taking into account that all quinolones have a common skeleton, it is logical to assume that the chelation between these compounds and Tb (III) ions will take place in a similar way. It has been suggested that the chelation occurs through the α-keto acid skeleton of the quinolone nucleus.

Another group of compounds that can form luminescent terbium chelates are salicylic acid and its substituted derivatives. One problem in the formation of the chelates between Tb (III) and these compounds is that they are only formed in alkaline media, and terbium emission is unstable due to the tendency of Tb (III) ions to form hydroxides at high pH media. This handicap has been solved by the formation of a ternary complex between terbium, EDTA and salicylic acid, but the presence of surfactants is commonly required to obtain the required sensitivity for the analytical applications. Taking into account the previous aspects, most of the research has been carried out in homogenous solution. However, the employment of the solid support has been recently introduced as another approach for the TSL detection of salicylic acid avoiding the use of highly alkaline media or surfactants [47]. By means of a

SIA optosensor, the determination of salicylic acid was satisfactorily performed in pharmaceuticals available in the Spanish Pharmacopoeia (manifold explained in next section).

The complexation between the analyte and the lanthanide ion can be performed on-line by direct confluence of solutions or off-line, adding the required volume of lanthanide ion solution to the sample flask and inserting directly the formed complex, which was the preferred option in the early optosensors. Tetracyclines [48] and anthracyclines [49] have been determined by europium-sensitized luminescence. The solid supports used were Chelex-100 and Amberlite XAD-2, respectively. The analytes were quantified in pharmaceutical and urine samples and the complex desorbed from the solid support using an HCl solution.

7.3. REFLECTOMETRY

Reflectance measurements are a convenient alternative to transmittance measurements in SPS as they circumvent the problems regarding the transparency of the solid support as well as the length of the optical path. This approach usually employs a bifurcated optical fiber bundle (one arm to conduct the radiation from the light source to the sample and other to conduct the radiation diffusively reflected towards the detector). This provides higher flexibility as the optical fiber permits to place the flow cell outside of the spectrophotometer. In addition, multiparametric measurements can be straightforwardly implemented, as several flow cells with solid phases selective to different analytes can be incorporated to the flow manifold.

Recently, many kinds of disk type sorbents have become available, and some attempts to use these membrane disks as the location for both the concentration of target colored chemicals and the direct measurement of the intensity of reflected incident light at the surface of the membrane disk have been carried out. The analytical method using the attenuation of intensities between the incident light and the reflected light, which is correlated with the amount of target chemical species concentrated on the membrane disk, is known as solid phase reflectometry. Miró et al. [50] used a laboratory-made flow-through sandwich cell for disk-based solid phase preconcentration and reflectance measurement. In comparison with the FI-SPS using the adsorbent beads packed in the flow-through cell, smaller particle sizes of adsorbent beads tightly bound within a PTFE support of the membrane disk are used.

Although these small adsorbents will give rise to a great back pressure and cannot be applied to a usual FI-SPS, the advantages of this reflectometry are due to such a small particle size, because of the high specific surface area; these small adsorbent beads show a higher capacity per unit weight and a higher adsorption rate for the analyte species. On the other hand, the light path length of a solid phase of 0.5 mm causes the sensitivity of this reflectometry to be inferior to that of FI-SPS.

Solid phases have been used to retain a product formed in solution in order to improve sensitivity. In this aspect, nitrite was determined in waters by injecting 2500 µL of the sample solution into a Shinn reagent carrier stream [50]. The product formed was retained by C_{18} silica gel coupled to an open sandwich-shape flow cell furnished with a bifurcated optical fiber for reflectance measurements. The product adsorbed onto the C_{18} disk was removed by injecting 150 µL of a 80% (v/v) methanol aqueous solution.

PVC membranes plasticized with 2-nitrophenyl-octyl-ether have been employed for determination of perchlorate ions in water [51]. The membrane was used to immobilize the lipophilic pH indicator 5-octadecanoyloxy-2-(4-nitrophenylazo)phenol and the ionic additive methyltridodecylammonium chloride, being the response mechanism based on the co-extraction of perchlorate and proton from a buffered solution of the anion. The extracted hydrogen ion then reacts with the pH indicator, generating a signal proportional to the perchlorate concentration. The optosensor presented a severe interference of bicarbonate ion, which would be eliminated by acidifying the sample previously the addition of the buffer solution.

Finally, a sensor employing a pH optical transduction was described for the determination of the pesticides carbaryl and propoxur in vegetables [52]. A controlled pore glass was used to covalently immobilize the enzyme acetylcholinesterase, whose activity in the hydrolysis of the acetylcholine to produce choline and acetic acid is inhibited by the pesticides.

7.4. VIBRATIONAL SPECTROSCOPY

Although vibrational spectroscopic detection does not present the sensitivity observed in most of the sensors developed employing luminescence detection, it provides structural information. The construction of systems based on Fourier-transform infrared spectroscopy (FTIR) detection is complex because the path length of the flow cell must be kept very short due to the high absorption of water in IR range. Thus, the extraction of the analyte from the aqueous solution seems to be a valuable alternative, as this procedure avoids the interference of water, besides pre-concentrating the analyte in the solid phase. In addition, once the extraction is performed, the interfering species can be separated from the analyte, improving the selectivity. Early procedures employed a SPE cartridge to pre-concentrate the analyte, followed by its elution with an appropriate organic solvent towards the detection cell [53].

There are not many middle IR (mid-IR) spectroscopy reports for analytical purposes. The combination with SPS is expected to be a very effective way to enhance the sensitivity. The first flow-through sensor was constructed by using a conventional mid-IR transmission cell furnished with a 55 μm polymeric spacer modified in order to hold Sephadex DEAE A-25 anion exchange resin beads, which swelled in contact with water, forming a gel-like disk between the two CaF_2 windows [54]. The optosensor was applied to the determination of acetic and malic acids in aqueous solutions by SIA. The addition of NaOH to the sample solutions produced the corresponding anions, allowing their retention by the anion exchange resin.

A mid-IR FTO for determination of carbohydrates in beer was also constructed by immobilizing amyloglucosidase on agarose beads [55]. Maltose standard solutions were automatically prepared in the SIA system and injected towards the detector. Once the solution reached the optosensor, the flow was stopped and spectra were run for 10 min in order to monitor the formation of the product. A water carrier stream was employed to flush the system and the content of carbohydrates in beer samples, expressed as maltose, was successfully determined by standard addition method.

Near infrared (NIR) was employed as detection technique in SPS for the determination of α-naphthylamine in water samples [56]. A commercial cell of 1 mm pathlength was filled with a C_{18}-bonded silica gel solid phase, which was conditioned/regenerated with 1 mL of a 20% (v/v) methanol solution.

Recently, an attenuated total reflection (ATR) based flow-through sensor has been reported for SPE and IR detection in a SIA system [57]. In this method, commercially available sorbent beads, LiChrolut EN (a polystyrene-divinylbenzene-based polymer) was used as the adsorbent for caffeine in soft drinks. A small amount of the adsorbent was packed into a diamond ATR flow cell. Similarly, Armenta and Lend reported the combination of FI-SPS and FT-IR [58]. Their target was also caffeine, but they employed C_{18} silica beads as the adsorbent.

The use of NIR detection allows working with glass cells and longer path lengths, but the superposition of overtones and combination bands in this spectral region causes a lower structural selectivity for NIR spectra. In this context, Raman spectroscopy appears as a very interesting alternative since water is a weak Raman scatterer and conventional glass flow-cells can be used to construct FTO. The determination of caffeine in energy drinks by a FT-Raman FTO based on a C_{18} solid phase was also described in the literature [59]. An anionic solid-phase reactor located before the C_{18} column was used to avoid matrix interference, improving the selectivity of Raman measurements. Sephadex QAE A-25 anion exchanger gel has also been used to determination of sulfonamides in pharmaceutical preparations by FT-Raman. The molecular and structural information contained in Raman spectra together with the selective retention of the species of interest on the anion-exchange sorbent made this method highly selective. The method was able to quantitate sulfathiazole and sulfamethoxazole in pharmaceutical preparations, avoiding interferences from other co-existing active principles [60].

Chapter 8

APPLICATIONS

The applications of FTO have been divided according to different compounds: pharmaceuticals and biological samples, food, trace elements, and pesticides.

8.1. PHARMACEUTICALS AND BIOLOGICAL SAMPLES

Most of the developed sensors by using this methodology are based on the direct measurement of the intrinsic UV absorbance or native fluorescence of the analyte. All the mono-parameter sensors with photometric detection are detailed in Table 1, including their detection limits (DL).

Three of the sensors were applied to the determination of inorganic elements in pharmaceuticals and biological fluids. In these sensors, a colorimetric organic compound was entrapped in the sol-gel inert support in order to form a complex with the analyte when arriving to the flow-cell. This way, copper, bismuth and zinc were satisfactorily determined [35,76,34].

Other examples include the spectrophotometric determination of soluble vitamins, such as ascorbic acid [20] or thiamine [61] in pharmaceutical preparations. However, when the targeted species does not present intrinsic UV absorbance, a chemical reaction can be used previously to the spectroscopy measurements. The determination of ascorbic acid based on the decrease of absorbance obtained when Prussian Blue was reduced by the analyte [72] or the quantitation of the phenothiazines promethazine and trifluoperazine after oxidation by Fe(III) and posterior complexation between Fe(II) and ferrozine [74] are two examples of this approach. In both cases, BI methodology was implemented in the system for solid support renewal.

In the case of multi-parameter sensors applied to clinical analysis (Table 2), most of them have been developed by means of the intrinsic absorbance of the analytes. The separation of the analytes can be made using a mini-column filled with the same solid support used in the flow-through cell or increasing the amount of solid support in the same flow-through cell.

Table 1. Mono-parameter FTO with photometric detection

Analyte	Flow method.	Solid support	Pre-treatment	DL ($\mu g\ mL^{-1}$)	Ref.
Thiamine	FIA	CM C-25	-	0.16	61
Ascorbic acid	FIA	QAE A-25	-	0.02	20
Diclofenac	FIA	QAE A-25	-	0.13	62
Minoxidil	FIA	SP C-25	-	0.006	63
Amoxycillin	FIA	QAE A-25	-	0.12	64
Pyridoxine	FIA	SP C-25	-	0.02	65
Paracetamol	FIA	QAE A-25	-	0.022	21
Adrenaline	FIA	QAE A-25	-	0.17	66
Tetracyclines	FIA	QAE A-25	-	0.07-0.12	67
Salicylic acid	FIA	QAE A-25	Acetylsalicylic acid also determined after hydrolysis	0.064	68
Zn	FIA	C_{18}	Chromogenic reagent immobilized on beads	0.01	69
Sulfonamides	FIA	QAE A-25	-	0.1	70
Ascorbic acid or Iron	FIA	QAE A-25	Complex with ferrozine; BIS	0.02 / 0.003	17
Cobalt	FIA	Dowex 50 W	Complex with 1-(2-pyridylazo)-2-naphthol; BIS	0.019	71
Ascorbic acid	FIA	QAE A-25	Reduction of Prussian Blue; BIS	0.08	72
Ciprofloxacin		SP C-25	-	0.035	73

Analyte	Flow method	Solid support	Separation procedure	DL	Ref.
Phenothiazines	FIA	QAE A-25	Oxidation by Fe (III), followed by Fe (II) and ferrozine complexation; BIS	0.1	74
Methylxanthines	FIA	C_{18}	-	0.09	75
Bismuth	MCFIA	Sol-gel	Complex with xylenol orange	0.007	76
Copper	MCFIA	Sol-gel	Complex with 4-(2-pyridylazo) resorcionol	0.003	35
Zinc	MCFIA	Sol-gel	Complex with 4-(2-pyridylazo) resorcionol	0.002	34
Acetazolamide	MCFIA	Sol-gel	Enzymatic measurements	44.5	77

Table 2. Multi-parameter FTO with photometric detection

Analyte	Flow method.	Solid support	Separation procedure	DL ($\mu g\ mL^{-1}$)	Ref.
Ascorbic acid Paracetamol	FIA	QAE A-25	Carrier solutions: different pH values	0.045 0.018	78
Paracetamol Salicylamide	FIA	QAE A-25	Carrier solutions: different pH values	0.1 0.24	79
Pyridoxine Thiamine	FIA	SP C-25	Minicolumn	0.084 0.1	13
Caffeine Dimenhydrinate Paracetamol	FIA	C_{18}	PLS	- - -	33
Caffeine Acetylsalicylic acid Paracetamol	FIA	C18	PLS	- - -	32

Table 2. (Continued)

Analyte	Flow method.	Solid support	Separation procedure	DL ($\mu g\ mL^{-1}$)	Ref.
Acetaminophen	FIA	C_{18}	Minicolumn	0.5	80
Acetylsalicylic acid				0.8	
Caffeine				0.3	
Ascorbic acid	FIA	SP C-25	Separation on the flow-cell	7	81
Acetylsalicylic acid				100	
Thiamine				0.75	
Ascorbic acid	FIA	QAE A-25	Two different flow cells	0.36	26
Thiamine		SP C-25		0.14	
Caffeine	FIA	C_{18}	Minicolumn	0.56	82
Paracetamol				0.75	
Caffeine	FIA	C_{18}	Minicolumn	0.65	83
Paracetamol				7.5	
Propyphenazone				1.9	
Copper	FIA	QAE A-25	chromogenic reagent Zincon; BIS	0.029	84
Zinc				0.04	
Sulfamethoxazole	FIA	SP C-25	Minicolumn	9.5	30
Trimethoprim				0.6	
Salicylamide	MCFIA	C_{18}	Minicolumn	0.33	85
Caffeine				0.15	
Piroxicam	MCFIA	C_{18}	Minicolumn	0.27	86
Pirydoxine				1.2	
Caffeine	MCFIA	C_{18}	Minicolumn	0.21	87
Salicylamide				0.61	
Propyphenazone				0.3	

In Figure 40, the typical manifold employing a minicolumn for the separation of two analytes is depicted. The sample is introduced by activating valves V_1 and V_2; one of the analytes is retained on the solid support of the mini-column while the other one passes through this and is retained on the solid support in the flow-cell. This first analyte develops its analytical signal on the sensing zone and is eluted from the microbeads by the carrier itself. After that, valves V_1 and V_3 are activated and the second analyte is eluted by an appropriate eluting solution from the mini-column and propelled towards the sensing zone, where it develops its corresponding analytical signal. Valve V_4 is used in the final cleaning step. This system has been used for the spectrophotometric determination of pirydoxine-piroxicam [86] and salicylamide-caffeine [85] in pharmaceutical preparations.

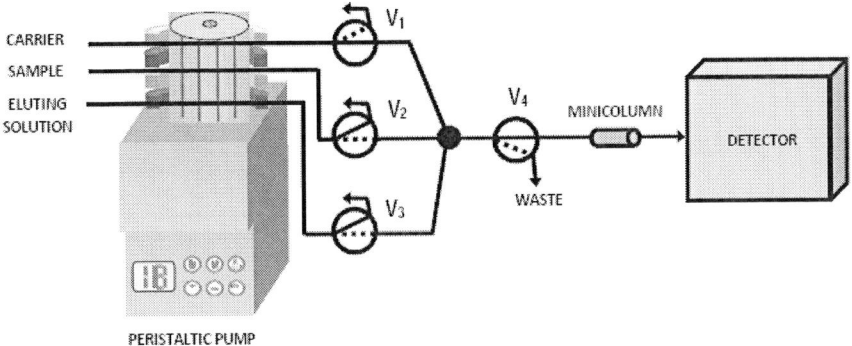

Figure 40. Manifold for analysis of pirydoxine-piroxicam or salicylamide-caffeine.

Mono-parameter applications in fluorometric sensors are shown in Table 3. Some examples include the determination of the drugs amiloride [89] and triamterene [91]. The implementation of PIF in FTO was described in 2005 for the determination of thiamine [39] in pharmaceuticals and human biological fluids. Other recent methods have been applied to the determination of several drugs, such as reserpine [96] or flufenamic acid [100].

Multi-parameter applications in fluorometric sensors are shown in Table 4. Manifold explained in Figure 40 can also be used for the fluorometric quantitation of furosemide and triamterene [104]. On the other hand, an additional solid support placed just above the detection area in the same cell has been employed for the fluorometric determination of naproxen-salicylic acid [106] in biological fluids and pyridoxine-riboflavin [105] in pharmaceuticals.

Table 3. Mono-parameter FTO with fluorometric detection

Analyte	Flow method.	Solid support	Pre-treatment	DL ($\mu g\ mL^{-1}$)	Ref.
Pyridoxine	FIA	SP C-25	-	0.00033	22
Pyridoxal	FIA	C_{18}	Reaction with Be	0.033	24
Riboflavin	FIA	C_{18}	-	0.0004	23
Dipyridamole	FIA	QAE A-25	-	0.00094	88
Amiloride	FIA	SP C-25	-	0.00033	89
Nafronyl or Naproxen	FIA	Silica gel Davisil or Amberlite XAD-7	-	0.021 0.013	90
Triamterene	FIA	SP C-25	-	0.00017	91
Propanolol	FIA	Amberlite XAD-7	-	0.001	92
Diphenhydramine	FIA	G-15	-	0.019	93
Quinidine	FIA	SP C-25	-	0.0002	94
Naphazoline	FIA	Amberlite XAD-7	-	0.0026	36
Thiamine	FIA	C_{18}	PIF	2.8×10^{-5}	39
Vanadium	FIA	QAE A-25	Complex with Alizarin Red S; BI	0.00045	95
Reserpine	FIA	C_{18}	PIF	5×10^{-5}	96
Digoxin	FIA	MIP		1.7×10^{-5}	97
Labetalol	SIA	C_{18}		0.0033	98
Paracetamol	SIA	QAE A-25	Reaction with $NaNO_2$	2	99
Flufenamic acid	MCFIA	C_{18}	PIF	0.00015	100
Digoxin	FIA	MIP	-	1.7×10^{-5}	38
Tetracyclines	FIA	Silica gel	-	0.0018	37
Warfarin	FIA	QAE A-25	-	0.0041	101

Table 4. Multi-parameter FTO with fluorometric detection

Analyte	Flow Method.	Solid support	Pre-treatment	DL ($\mu g\ mL^{-1}$)	Ref.
Salicylamide Salicylic acid	FIA	QAE A-25	pH changes determine which analyte is retained	0.0009 0.002	102
Pyridoxal Pyridoxic acid Pyridoxal-5 phosphate	FIA	C_{18}	Reaction with Be	0.0084^b 0.0092^b 0.012^b	103
Sulfanilamide Sulfamethoxazole Sulfathiazole[a]	FIA	QAE A-25	PIF	0.0029 0.0081 0.0057	40
Furosemide Triamterene	MCFIA	SP C-25	Separation on precolumn	0.015 0.0001	104
Pyridoxine Riboflavine	MCFIA	C_{18}	Separation on the flow-cell	0.045 0.003	105
Naproxen Salicylic acid	MCFIA	C_{18}	Separation on the flow-cell	0.0003 0.0013	106
Ascorbic acid Pyridoxine Riboflavine	SIA	QAE A-25 C_{18}	Reaction with MnO_4^- on beads. Separation on the flow-cell	9.1 0.12 0.008	25

[a]Binary mixtures of sulfanilamide / sulfamethoxazole or sulfanilamide / sulfathiazole; [b]Quantitation limit

A recent paper demonstrated the versatility of SIA for the design of complex automatic optosensors. The method, employed for the determination of vitamins B_2, B_6 and C [25], presented a novel characteristic: the use of two different detection techniques in the same system, therefore enhancing the scope of applications. This approach allowed the determination of compounds with dissimilar spectroscopic characteristics. Manifold is depicted in Figure 41. By means of the SIA multivalve, the sample could be easily pumped towards the required detector, therefore allowing the use of the double-detection technique. Taking into account the native fluorescence of B_6 and B_2, these analytes were determined by means of a spectrofluorometer. However, as ascorbic acid does not present native fluorescence, its reaction with permanganate in a sulphuric medium was performed and the obtained CL measured. In every case, the determination was based on the direct measurement of the analytical signal of the respective vitamin retained on the sensing solid support placed on the respective detection area. Two different solid supports were used, a non-ionic C_{18} silica gel for the fluorescent compounds and an anionic QAE-A25 for ascorbic acid.

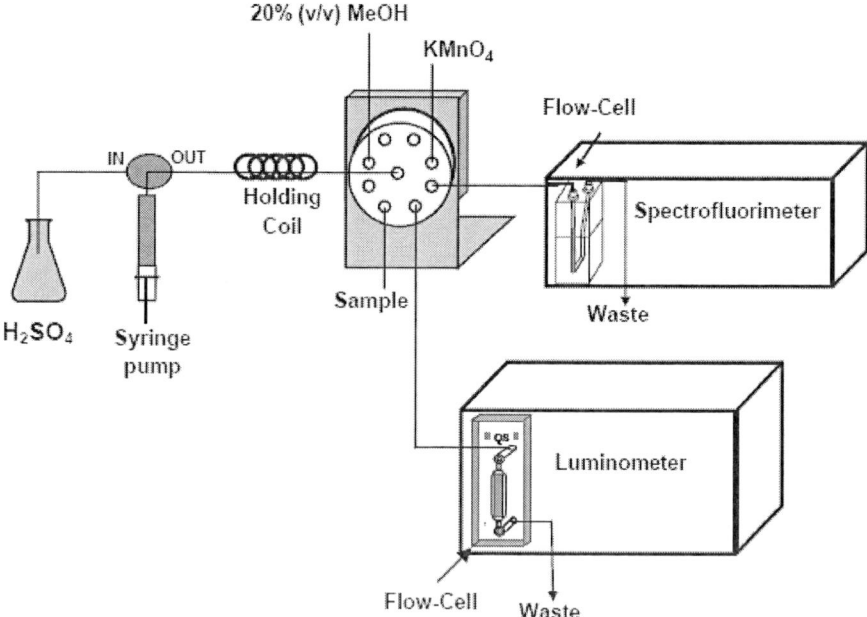

Figure 41. Manifold for the determination of vitamins B_2, B_6 and C.

Table 5. FTO with other luminescence detection

Analyte	Flow method.	Solid support	Pre-treatment	DL ($\mu g\ mL^{-1}$)	Detect.	Ref.
Tetracyclines	FIA	Amberlite XAD-2	Complex with Eu (III)	0.0002-0.0004	LSL	48
Antracyclines	FIA	Amberlite XAD-2	Complex with Eu (III)	10^{-8}M	LSL	49
Analgin	FIA	Cation exchanger 732	Reaction with acidic rhodamine 6G	0.15	CL	45
Ciprofloxacin	FIA	Amberlite IRC-50S	Complex with Eu (III)	0.033	LSL	107
Ofloxacin	FIA	CR-1211 sponge	Oxidation with PbO_2	0.078	CL	108
Pipemidic acid	FIA	CR-1211 sponge	Oxidation with $NaBiO_3$	0.062	CL	109
Reserpine	FIA	Amberlyst A-27	Oxidation between luminol and periodate	0.0003	CL	110
Analgin	FIA	CR-1211 sponge	Oxidation with MnO_2	27	CL	111
Naphazoline	FIA	Amberlite XAD-7	Use of KI	0.0094	RTP	42
Norfloxacin	FIA	SP C-25	Complex with Tb (III)	0.0015	LSL	112
Salbutamol	FIA	MIP	Luminol and potassium ferricyanide reaction	0.016	CL	113

Table 5. (Continued)

Norfloxacin	FIA	AG 1-X8	KMnO$_4$ and Na$_2$SO$_4$ reaction	CL	0.0028	LSL	114
p-aminobenzoic acid	FIA	QAE A-25	Complex with Tb (III)		0.06	LSL	115
Fenfluramine	FIA	MIP	Oxidation with KMnO$_4$		1	CL	116
Salicylic acid	SIA	QAE A-25	Complex with Tb (III) on beads		0.045	LSL	47
Cromolyn	SIA	Chelex-100	Complex with Tb (III)		0.015	LSL	117
Pipemidic acid	MCFIA	SP C-25	Complex with Tb (III)		1.79×10^{-7}	LSL	118
Salicylic acid	MCFIA	QAE A-25	Reaction with MnO$_4^-$ on beads		0.3	CL	119
5-Aminosalicylic acid	MCFIA	QAE A-25	Reaction with MnO$_4^-$ on beads		0.3	CL	120
Amoxicillin	FIA	MIP	Reaction with MnO$_4^-$		0.0013	CL	46
Cefadroxil	MCFIA	QAE A-25	Reaction with MnO$_4^-$ on beads		0.29	CL	121
Orbifloxacin	SIA	CM C-25	Complex with Tb (III) off-line		0.0033	LSL	122

Sensors with other luminescence detection different to fluorescence are depicted in Table 5. CL has been used as detection technique using two main approaches:

(1) *To place the solid support inside the flow-cell*, which is situated in front of the window of the PMT. It has been the most commonly employed. In this case two options are available: a) both reagents and analyte solutions can be delivered towards the cell; b) the reagents can be previously immobilized on the solid support and only the sample solution is inserted for each determination. One example of the first case is depicted in Figure 42A for the determination of salbutamol [113]. This method is based on the sensitization produced by the analyte on the CL reaction between luminol and ferricyanide potassium. A MIP was employed as solid support and all solutions pumped towards the cell for each sample determination. The second alternative would be to retain the oxidant on the solid support previously to the insertion of the analyte, which is the only solution pumped towards the flow-cell, where the reaction takes place. This approach has been used for the determination of ofloxacin using PbO_2 retained on a sponge rubber [108] or for the quantitation of pipemidic acid based on its sensitizing effect on the CL oxidation of sulphite by sodium bismuthate, which was previously immobilized on the solid support inside the flow-cell [109].

(2) *To immobilize the reagents on a solid support placed just before the cell*, releasing then in the appropriate moment to obtain the CL reaction with the analyte. This approach was applied to the determination of reserpine in pharmaceutical preparations and human urine samples [110]. The CL reagents, luminol and periodate, used in this sensor, were both immobilized on anion-exchange resin. Through injection of 100 μL eluting solution, the reagents on the anion-exchange resin column were eluted and in the presence of reserpine, the CL intensity was decreased, by which reserpine could be sensed. The sensor showed stability without replacing the resin microbeads for at least 80 h. Manifold employed in this method is depicted in Figure 42B.

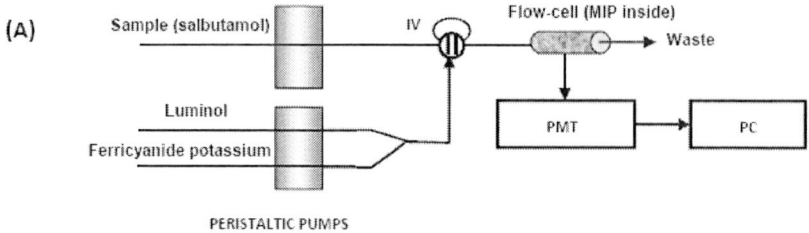

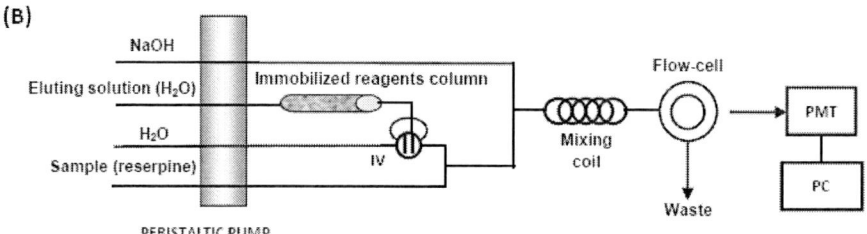

Figure 42. Manifold for the determination of: (A) salbutamol; (B) reserpine.

One paper has been published comparing different phosphorescence methodologies for determining naphazoline in pharmaceuticals [42]. One of the possibilities consisted on an optical sensor, employing iodide as heavy atom and Amberlite XAD-7 as the solid support. The sensitivity and repeatability obtained with the optosensor was similar to the ones observed in solution.

LSL has been used in several optosensors, employing Eu (III) or Tb (III) as the lanthanide ion. In all cases, high sensitivity and selectivity was observed due not only to the solid support, but also to the highly specific detection technique. Different flow methodologies have been used with LSL detection (FIA, SIA and MCFIA), which will be described below.

Coupling FIA-LSL. Two different methods have been developed for the determination of the quinolone norfloxacin. In the first method, the chelate between Tb (III) and this analyte is formed on-line and retained on a cation-exchanger resin (Sephadex SP C-25) placed in the flow-through cell, where norfloxacin excitation and terbium ion emission take place [112]. Manifold is shown in Figure 43 and the procedure is as follows: the sample solution (with acetate buffer solution 0.1 M at pH 5.6) containing norfloxacin confluences with terbium 4×10^{-3} M solution previous to the sample loop, so the complex between them is formed. After the confluence, this mixture fills the sample loop, which is then inserted into the carrier stream by using the six-port rotary injection valve (IV). No reactor is needed for the formation of the complex, due to the fast chelation. The complex is transported by the carrier towards the flow-through cell, where it develops its luminescence signal on the solid support. Once the signal has achieved its maximum, by turning the selection valve (SV), a 0.08 M EDTA eluting solution is injected in order to remove the complex from the resin, regenerating the solid microbeads.

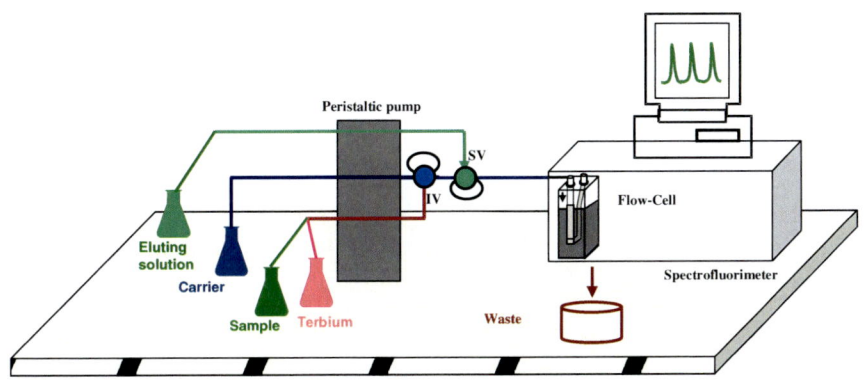

Figure 43. Manifold for the determination of norfloxacin (I).

The second system developed for determining norfloxacin was based also on the emission intensity from the Tb (III) solution sensitized by this analyte [114]. The excitation of the organic compound was not achieved by the excitation light of a spectrofluorometer, but by means of a CL reaction. An anion-exchanger resin, AG 1-X8, was packed in a glass column, and permanganate ions immobilized on the microbeads. When sodium sulphite is introduced in the system, its CL reaction with permanganate takes place and excited sulfur dioxide is produced (SO_2^*). Then, an energy transfer from SO_2^* to norfloxacin occurs, producing excited norfloxacin molecules. Finally, a chelate between norfloxacin* and Tb (III) ion is formed and the energy transfer from the organic compound to terbium is produced, measuring its emission. In this case, a spectrofluorometer is not needed, but just a PMT to measure the analytical signal. The diagram of the flow system is depicted in Figure 44.

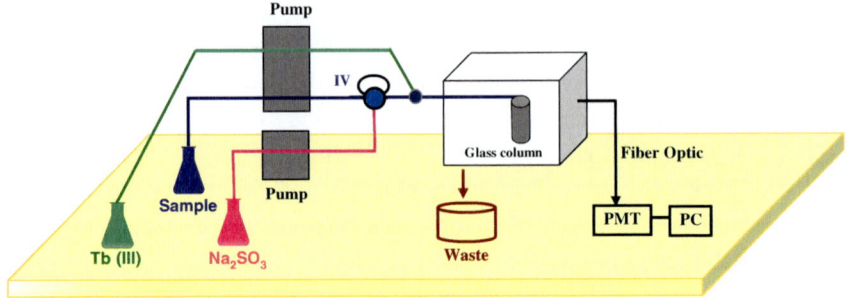

Figure 44. Manifold for the determination of norfloxacin (II).

Coupling SIA-LSL. Three TSL optosensors have been developed making use of this methodology, for determining salicylic acid, cromolyn and orbifloxacin. Manifold used for the determination of salicylic acid and cromolyn is depicted in Figure 45.

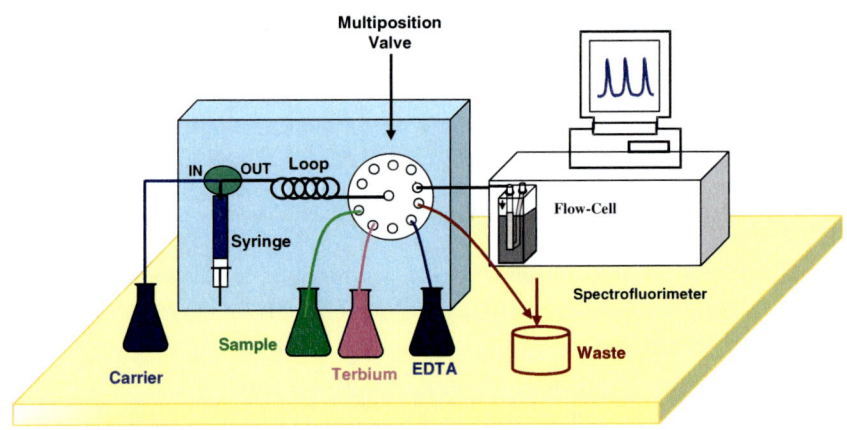

Figure 45. Manifold for the determination of salicylic acid and cromolyn

The procedure employed for the determination of salicylic acid [47] in pharmaceuticals from the Spanish Pharmacopoeia is as follows: (1) before starting the measuring process, the carrier, deionized water, was passed through the solid support in order to condition it. An anion-exchanger, Sephadex QAE A-25, was used to retain the anionic analyte; (2) the sample containing the analyte was aspirated and pumped toward the flow-through cell, remaining salicylic acid retained on the sensing zone; (3) after that, and separated by a small volume of carrier to avoid the mixing of sample and terbium solutions, 2×10^{-3} M Tb (III) solution was also aspirated and pumped toward the detection zone; (4) with the arrival of Tb (III) to the sensing zone, the reaction between the analyte and terbium takes place and the analytical signal is obtained at 300/545 nm ($\lambda_{ex}/\lambda_{em}$); and (5) after obtaining the signal, the support was regenerated with 0.08 M EDTA solution. This last step produced a small shoulder at the end of each signal peak.

A typical profile of the signal is shown in Figure 46. All the results obtained were in agreement with those provided by the manufacturers of the pharmaceuticals.

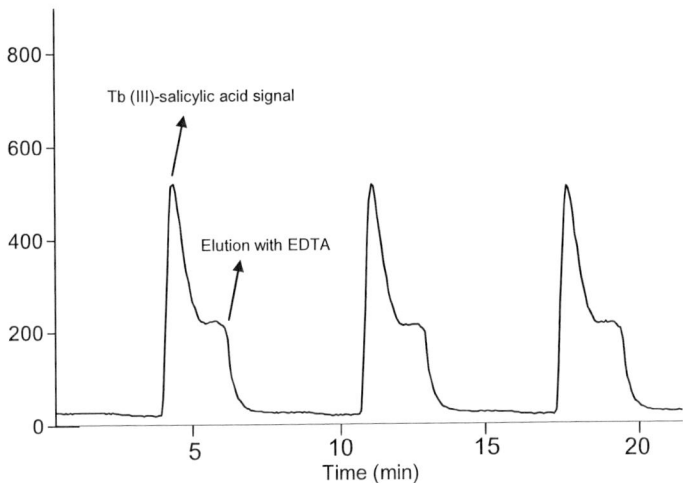

Figure 46. Flow profile of the signal obtained for salicylic acid.

The same manifold was used for the determination of cromolyn [117], but the procedure was slightly different: (1) the solid support (cationic-exchanger Chelex-100) was conditioned when the flow cell was filled for the first time with the carrier solution, 0.05 M sodium acetate/acetic acid solution, pH 5.9; (2) two aliquots of sample solution and another two aliquots of 6×10^{-3} M Tb (III) solution were sequentially aspirated, in double-sandwich mode, into the syringe pump, using a higher volume of solutions when human urine was analyzed in order to obtain higher sensitivity; (3) carrier solution was also aspirated and the mix was pumped towards the flow cell passing previously through the holding coil, where the Tb (III)-cromolyn chelate was formed; and (4) once the chelate arrived to the sensing zone and its analytical signal was registered at 336/545 nm ($\lambda_{ex}/\lambda_{em}$), the solid support was regenerated by passing through it a 0.05 M EDTA eluting solution. In all cases, the recoveries were close to 100%, showing the suitability of this method for the routine analysis of the drug in clinical analysis.

In the case of the determination of orbifloxacin [122], the same SIA system was used, but the formation of Tb (III)-orbifloxacin complex was performed off-line. As a result, terbium solution was not aspirated through the valve and sample solution already included Tb (III). The procedure was as follows: (1) the conditioning of the solid support (cationic-exchanger Sephadex CM C-25) was performed with the carrier, acetate buffer 0.05 mol L^{-1}, pH 6; (2) carrier and sample were sequentially aspirated into the syringe pump and holding coil; (3) the sample solution was pumped toward the flow-through cell, where the chelate was retained on the sensing zone, measuring its analytical signal at 275/545 nm; and (4) once the signal from the chelate was recorded, a 0.08 mol L^{-1} EDTA eluting solution was used to regenerate the solid support. This method was applied to the determination of the drug in dog and horse urine.

Coupling MCFIA-LSL. A manifold comprising five 3-way solenoid valves was developed for the determination of the quinolone pipemidic acid [118]. This manifold is depicted in Figure 47 as well as the complete timing procedure of the valves. The procedure is the following: (1) initially, all valves are switched off and the carrier, 0.15 M acetic acid/sodium acetate buffer, pH 5.6, flows through the flow cell while all other solutions are recycling to their respective vessels; (2) the sample and terbium solutions are introduced and mixed by simultaneously switching the valves V_1, V_2 and V_3 on for the required time. This way, the carrier solution is recycled while the Tb-analyte chelate is being formed on-line; (3) all valves are switched off, being the chelate carried towards the sensing microbeads (cation-exchanger Sephadex SP C-25) by the carrier solution. The chelate develops its transitory luminescent signal in the flow-through cell; (4) after the maximum is reached and a partial elution of the complex is observed, valves V_1 and V_4 are switched on, and the 0.08 M EDTA eluting solution is introduced into the flowing system, regenerating the sensing support; (5) and finally, the portion of tubing still filled with sample solution is cleaned with the next sample solution in order to avoid any possible contamination between samples. In the timing procedure of the valves (bottom part of Figure 47), $T_{1, 2...}$ refer to the time course during which solenoid valves $V_{1, 2...}$ were switched on and off. The filled rectangles above the valve's time line indicate the time at which the corresponding valves were switched on. The steps were as follows: 1: sample introduction; 2: signal development; 3: elution; 4: cleaning step.

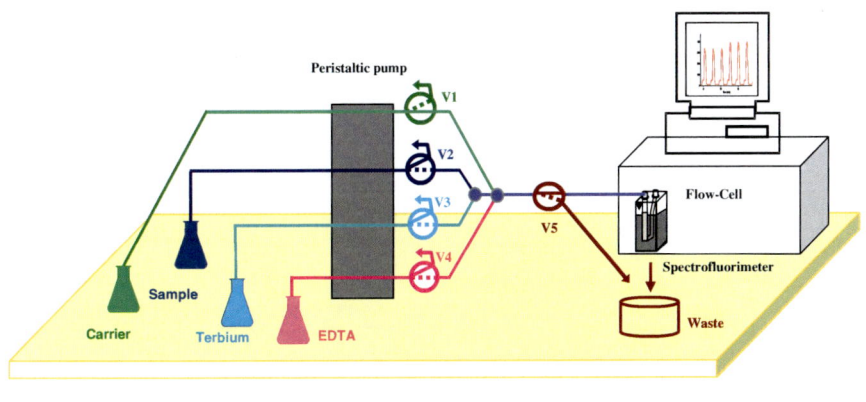

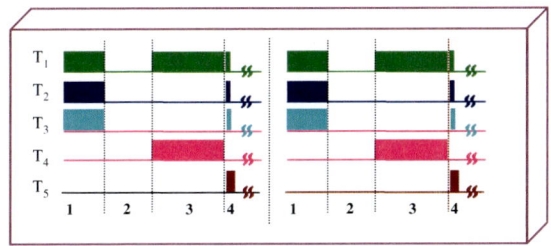

Figure 47. Upper part: Manifold used for determination of pipemidic acid. Bottom part: Valves scheme.

8.2. FOOD

FTO is also applicable to the analysis of organic or inorganic chemical components in food samples. Capitán-Vallvey and co-workers successfully carried out the online separation of analytes in a binary mixture in a flow-through cell (Hellma 138-QS) packed with the appropriate adsorbent. The light absorption caused by one analyte with a shorter retention time could be first separately measured, while the other is transiently retained in the upper part of the solid particle layer in the flow cell where the incident light did not enter. The proposed system was used for the simultaneous determinations of butylated hydroxyanisole and n-propyl gallate [123], and saccharin and in sweets and drinks [124].

Diphenylamine can be determined in apple and pear by a fluorometric FI-SPS [125] after its preconcentration online on C_{18} silica beads. On the other hand, a HAI-RTP method was applied to the screening of nafcillin in milk-based product [15] using a MIP for its selective adsorption in the flow cell.

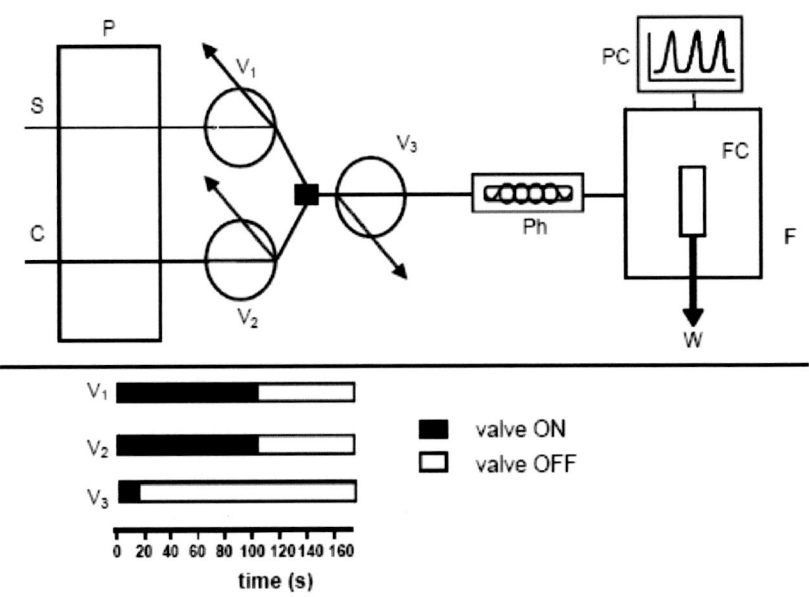

Figure 48. Manifold for analysis of piceid.

A PIF determination of piceid in cocoa-containing products with the aid of a multicommuted continuous-flow assembly, which was provided with an on-line photoreactor, was developed [126]. A strongly fluorescent photoproduct is generated from piceid when it is irradiated under UV light for 30 s, which is retained on Sephadex QAE A-25. The sample pre-treatment included delipidation with toluene and cyclohexane followed by stilbenes extraction with 80% (v/v) ethanol solution and clean-up by SPE on C_{18} cartridges. Manifold is shown in Figure 48.

Likewise, a novel method has been established for the determination of resveratrol [127], a phytoalexin that plays a central role in the human diet because of their antioxidant, anticarcinogenic and antimutagenic properties. A selective extraction of resveratrol from beer and posterior determination by PIF (277/382, $\lambda_{ex}/\lambda_{em}$) in a multicommutation assembly was developed. Sephadex QAE A-25 was used as solid support in the flow cell. The extraction and cleaning up procedure was carried out with C_{18} cartridges, by eluting resveratrol with methanol, which showed extraction efficiencies over 80%. The previous washing of cartridges with 40% (v/v) methanol solution allowed removing piceid and other interfering substances present in beer.

8.3. TRACE ELEMENTS

FTO in the field of trace elements are described in Table 6. They have been applied in most of the cases to water samples (drink, sea, tap water, etc.) by using FIA as flow methodology. In fluorometric optosensors, both the previous loading of the reagent on the solid support [137] and the on-line formation of the chelate before the flow-cell [135,136] strategies have been used. When the reaction takes place before the cell, sometimes it is necessary to use a reaction coil to fulfill the chelation. This strategy lowers the sensitivity of the system and decreases the sample throughput.

The analysis of environmental samples has focused not only in essential trace elements, such as vanadium or iron [129], but also in the analysis of toxic elements. For instance, there is a great demand for the determination of lead in environmental samples. This metal is now widespread, contaminating virtually the whole biosphere, and its determination has been performed using RTP detection [131].

On the other hand, chromium is mostly present in two different oxidation states, Cr (III) and Cr (VI). Cr (III) is an essential trace element, however, Cr (VI) is very toxic to organisms and it has been demonstrated that its compounds are carcinogenic to humans. An on-line oxidation method for determining chromium, with diphenylcarbazide as chromogenic agent, was developed using a flow electrolysis cell. This system could be applied to the total Cr determination together with the specific determination of Cr (III) and Cr (VI) at trace levels in river water samples [128].

Table 6. FTO for the analysis of trace elements

Analyte	Solid support	Pre-treatment	DL ($\mu g\ L^{-1}$)	Det.	Ref.
Chromium	Muromac 50W-X2	Reaction with diphenylcarbazide	0.014	PH	128
Iron	C_{18}	Reaction with 1-(2-thiazolylazo)-2-naphthol	15	PH	129
Iodide	Dowex 1x2-200	Reaction with chelate aluminium-quinolin-8-ol-5-sulphonic acid	10^a 5^a	RTP FL	130
Lead	Dowex 1x2-200	Reaction with derivatives of quinolinesulphonic acid	0.1	RTP	131
Mercury	Amberlite XAD-4	Reaction with thiamine	3	FL	132
Aluminium	Chelex 100	Reaction with quinolin-8-ol-5-sulfonate	$3\times10^{-9}M$	FL	133
Zinc	C_{18}	Reaction with p-(tosylamino)quinoline	0.9	FL	134
Cadmium	QAE A-25	Reaction with 8-hydroxyquinoline-5-sulfonic acid	0.48	FL	135
Aluminium	C_{18}	Reaction with chromotropic acid	2.6	FL	136
Aluminium Berillium	QAE A-25	Reaction with morin	0.01^a 0.024	FL	137

$^a\mu g\ mL^{-1}$; PH: spectrophotometry UV-Vis; FL: fluorometry

It is important to mention that the determination of more than one element can also be performed by modifying the flow system. In this way, aluminium (III) and beryllium (II) could be fluorometrically determined using a BIS–FIA system [137]. The sensor was based on the alternate use of two carrier solutions and the Hellma 176-QS flow cell. The cell was filled by injecting a homogeneous bead suspension of an appropriate solid support (Sephadex QAE A-25) previously loaded with the fluorogenic reagent morin. By using the first carrier solution, only the reaction between aluminium and morin is accomplished when the sample is inserted, obtaining the corresponding signal. After that, the carrier is changed and the sample solution inserted once more. This time, only beryllium reacts with morin. At the end of the analysis, beads are automatically discarded from the flow cell, by reversing the flow, and transported out of the system. Hence, both metals can be determined using the same flow system, just by changing the carrier solution. This multisensor was applied to the determination of these elements in tap and river waters.

8.4. PESTICIDES

Few FTO with photometric detection have been developed for the analysis of pesticides (Table 7). The use of mathematical treatments has been employed for the determination of the carbamates carbofuran, propoxur and carbaryl [138]. The reaction products obtained after hydrolysis and coupling to diazotized sulfanillic acid were retained on C_{18} beads and their signals monitored using a diode array spectrophotometer. The nine wavelengths selected for the simultaneous monitoring resulted in maximal differences between the absorption spectra of the three analytes. A calculation program (MIXFIA) based on the establishment of nine equations with three unknowns was used.

It is worthy mentioning the design of an optical fiber reflectance sensor for the quantification of 1-naphtylamine [139] making use of the flow methodology MSFIA. This analyte is one of the main degradation products of pesticides derived from naphthalene acid and a well-known carcinogen. The Griess reaction was used in this method in order to form an azo dye, which was preconcentrated and detected in-situ on C_{18} extraction disks.

The main detection technique used in FTO for pesticide determination is fluorescence (Table 8). An example of these systems is the optosensor developed for the screening of bitertanol in banana samples, in which the native fluorescence of the pesticide was monitored at 261/326 nm after its retention on C_{18} silica gel [150]. The sensor showed good tolerance to the presence of other common pesticides such as simazine, imazalil, dimethoate or aldicarb. Nevertheless, the interference of carbofuran and carbaryl was very serious and the authors proposed two changes in the procedure in order to their elimination. The proposed method fulfilled the maximum residue level established for bitertanol in banana by European Union, 3 mg kg^{-1}.

Table 7. FTO with photometric detection

Analyte	Solid support	Matrix	Principle	DL ($\mu g\ L^{-1}$)	Ref.
Carbaryl Propoxur	CPG	Onion; lettuce	pH indicator used as transducer of acetylcholinesterase inhibition	50 8	52
Carbofuran Propoxur Carbaryl	C_{18}	Waters	Derivatization procedure; chemometrics	500 500 500	138
1-naphthylamine	C_{18} extraction disks	Waters	Griess reaction	1.1	139
Paraquat	Dowex 50W-X8-200	Waters; soils	Reaction with dithionite ion	0.11	140

CPG: controlled-pore glass

In order to achieve the highest possible automation and lowest wastes generation, MCFIA has been also used in FTO for the development of monoparameter and biparameter optosensors. The well-known QuEChERS (Quick, Easy, Cheap, Effective, Rugged and Safe) extraction methodology [141], consisting of an acetonitrile extraction/partitioning and dispersive SPE clean up with primary secondary amine, was used in some of them. This methodology, allowed the determination of the target analytes in different food samples, such as apples, pears and citrus fruits [125,157]. In these cases, the DLs found were low enough to fulfill the maximum residue level regulated by the legislation.

Table 8. FTO with fluorometric detection

Analyte	Solid support	Matrix	Principle	DL (µg L^{-1})	Ref.
Atrazine	Protein A/G	Waters; vegetables	Polyclonal antisera	0.15	142
Atrazine; Diuron; Isoproturon; Paraquat; Simazine	Biochip membrane	---	Inhibition of algae photosynthesis; array sensor	0.5-100	143
Carbaryl	Protein A/G	Tomatoes; peppers	Monoclonal antibodies	2	144
Carbaryl	CPG	Waters; honey	Monoclonal antibodies	0.029	145
Isoproturon	CPG	Waters	Polyclonal antibodies; isothiocyanate as label	3	146
1-naphtol	Protein A/G	Waters	Polyclonal antibodies	12	147
Azoxystrobin	C$_{18}$	Musts; grapes; wines	PIF	2.4	148
Benomyl Thiabendazole	C$_{18}$	Waters; pharmaceuticals	Separation in flow-cell	3.6 0.06	149
Bitertanol	C$_{18}$	Bananas	QuEChERs	0.014^a	150
Carbaryl	MIP	Waters	Direct measurement	0.27	151
Carbendazim Carbofuran Benomyl	C$_{18}$	Waters	Separation in precolumn	15 68 35	27
Diphenylamine	C18	Apples; pears	QuEChERS	0.06a	125
Fuberidazole Carbaryl Benomyl	C18	Waters	Separation in precolumn	0.09 6 9	152

Table 8. (Continued)

Analyte	Solid support	Matrix	Principle	DL (µg L-1)	Ref.
Fuberidazole o-phenylphenol	C$_{18}$	Waters	Separation in flow-cell	0.18 6.1	29
Imidacloprid	C$_{18}$	Peppers; waters	PIF	1.8	41
Linuron	C$_{18}$	Waters	Micellar medium; PIF	130	153
Metsulfuron methyl	C$_{18}$	Waters	Micellar medium; PIF	0.14	154
α-naphtol o-phenylphenol Thiabendazole	C$_{18}$	Waters	Chemometrics	2 2 2	155
Simazine	CPGA	Waters	Antibodies	0.0013	156
Thiabendazole	C$_{18}$	Citrus fruits	QuEChERS	0.09^a	157
Thiabendazole	Nylon	Waters	-	2.8	3
Thiabendazole Metsulfuron methyl	C$_{18}$	Waters	Separation in minicolumn; micellar medium; PIF	2.5 3.3	28
Thiabendazole Warfarin	C$_{18}$	Pesticide preparations	Separation in precolumn	2.35 0.54	158
Warfarin	Cyclobond I	Waters	Complex between warfarin and β-cyclodextrin	19	159
Warfarin	Sephadex QAE A-25	Waters	Direct measurement	4.1	101

CPG: controlled-pore glass; HRP: horse radish peroxidase; ^{a}mg kg^{-1}

Table 9. FTO with other luminescence detection

Analyte	Solid support	Matrix	Principle	DL ($\mu g\ L^{-1}$)	Det.	Ref.
Aldicarb Paraoxon	Eupergit C	Soil; vegetables	Acetylcholinesterase inhibition measured with luminol reaction	4 0.75	CL	160
Atrazine Diuron	Magnetic beads	–	Photosystem II inhibition measured with luminol reaction	6.5 2.3	CL	161
Chloramphenicol	Biodyne B	Meat; milk	Antibody on the solid support; detection based on luminol reaction	3.23	CL	162
2, 4-D	MIP	–	Labelling with tobacco peroxidase	0.5	CL	163
Paraoxon[a]	Glass beads	–	Inhibition of alkaline phosphatase, which catalyzes the CL reaction	50	CL	164
Maleic hydrazide	MIP	Potatoes; onions	Reaction with luminol and potassium periodate	60	CL	165
Naptalam 1-naphthylamine	Amberlite XAD 7	Waters	$TlNO_3$ and Na_2SO_3 in carrier solution	8.1 11.2	RTP	166
Thiabendazole	Nylon	Waters	Potassium iodide and sodium sulphite in sample solution	12.9	RTP	167

2, 4-D: 2, 4-dichlorophenoxyacetic acid; [a]Model analyte for organophosphorus pesticides

The resolution of three pesticides, α-naphtol, thiabendazole and o-phenylphenol, at µg L^{-1} level was possible by developing a FIA-based system with fluorometric detection, using C_{18} silica gel as active sorbent substrate in the flow-cell [155]. For the resolution, it was necessary the use of a mathematical treatment of their analytical signals. Sensors with CL or RTP detection are shown in Table 9.

Maleic hydrazide was selectively immobilized on a MIP filling the flow-cell [165]. Then, luminol and potassium periodate solutions were delivered to the cell and originated strong CL. This approach did not provide a high throughput since the measurement involved: (a) immobilization of the analyte on the MIP, (b) washing of the MIP to remove the resting sample solution, (c) delivering of CL reagents to the flow-cell, and (d) washing of the MIP for its regeneration. Nevertheless, the MIP used as recognition material of maleic hydrazide allowed detecting this latter in vegetable samples without purification step.

Chapter 9

CONCLUSIONS AND TRENDS

The successful operation of a modern analytical laboratory requires accurate analysis of samples in the shortest time possible after receipt. Recent advances in instrumentation have led to increased interest in the development of automated chemical analyses. Flow systems have been widely developed in routine analysis as they allow fast analysis with a minimal sample handling in many cases. Furthermore, other advantages such as minimal reagent consumption in some applications and the possibility of adapting these systems in the field should be highlighted. These reasons make that the development and applications of chemical and biochemical FTO have shown fast growth over the past two decades.

Nowadays, the modern methods for organic compounds detection usually involve liquid or gas chromatography coupled to mass spectrometry detection (HPLC-MS, GC-MS). These methods present the advantages of high sensitivity, high selectivity and the possibility of multi-residue analysis [168]. However, the instruments involved are not available in every analytical laboratory and qualified operators are required due to the complexity of these methods of analysis. Although chromatographic methods with MS detection are needed for the development of multi-residue methods in specific fields of research, one of the most promising key areas in recent studies is the research efforts devoted to the development of FTO for simultaneous determination of several compounds (multioptosensors) which show both a much lower cost per analysis and a very much higher sampling rate.

FTO are useful in the routine analysis of pharmaceutical preparations, biological samples, foods or contaminants when only a few number of compounds need to be monitored. Most of them have been applied to the determination of the target compounds in real samples. Only few papers have dealt with determining a family of compounds. In general, these optical sensors were applied to the quantification of one, two or three specific compounds. Thus, the main handicap of these sensors is the impossibility to perform multi-residue analysis, as more than three pesticides have not been determined up to date in the same analytical method. Taking into account that complex samples such as fruits, vegetables, meat samples, biological fluids or soils have been already analyzed using these FTO, the main trends would be: (1) design of miniaturized instruments for in-situ measurements; (2) further investigation in order to determine several analytes in the same analysis.

The cited flow systems are characterized by their high sensitivity and selectivity, easy automation, less amount of reagents used and low cost of equipment. They allow to develop very simple and reliable methodologies that contribute to a greener analytical chemistry by saving sample, reagents, and waste generation. Although most of the solid supports used in the described systems are already available in the market, the availability of some of them (e.g. MIPs) still has to be increased. From the point of view of the author, FTO offer a very attractive, promising and fruitful research field on the development of small portable and automated systems, which could be applied to a high number of species simultaneously in real samples.

REFERENCES

[1] Yoshimura, K., Waki, H., & Ohashi, S. (1976). Ion-exchanger colorimetry-I. Microdetermination of chromium, iron, copper and cobalt in water. Talanta, 23, 449-454.

[2] Yoshimura, K. (1987). Implementation of ion-exchanger absorptiometric detection in flow analysis systems. Analytical Chemystry, 50, 2922-2924.

[3] Piccirilli, G.N., & Escandar, G.M. (2007). A novel flow-through fluorescence optosensor for the determination of thiabendazole. Analytica Chimica Acta, 601, 196-203.

[4] Molina Díaz, A., Ruiz Medina, A., & Fernández de Córdova, M.L. (2002). The potential of flow-through optosensors in pharmaceutical analysis. Journal of Pharmaceutical and Biomedical Analysis, 28, 399-419.

[5] Ruzicka, J., & Hansen, E.H. (1975). Flow injection analyses. Part I. A new concept of fast continuous flow analysis. Analytica Chimica Acta, 78, 145-157.

[6] Ruzicka, J., & Marshall, G.D. (1990). Sequential injection: A new concept for chemical sensors, process analysis and laboratory assays. Analytica Chimica Acta, 237, 329-343.

[7] Reis, B.F., Giné, M.F., Zagatto, E.A.G., Lima, J.L., & Lapa, R.A. (1994). Multicommutation in flow analysis. Part 1. Binary sampling: concepts, instrumentation and spectrophotometric determination of iron in plant digests. Analytica Chimica Acta, 293, 129-138.

[8] Molina-Díaz, A., García-Reyes, J.F., & Gilbert-López, B. (2010). Solid-phase spectroscopy from the point of view of green analytical chemistry. TrAC - Trends in Analytical Chemistry, 29, 654-666.

[9] Matsuoka, S., & Yoshimura, K. (2010). Recent trends in solid phase spectrometry: 2003-2009. A review. Analytica Chimica Acta, 664, 1-18.
[10] Yoshimura, K., & Waki, H. (1985). Ion-exchanger phase absorptiometry for trace analysis. Talanta, 32, 345-352.
[11] Fernández de Córdova, M.L., Molina Díaz, A., Pascual-Reguera, M.I., & Capitán Vallvey, L.F. (1995). Solid-phase spectrophotometric determination of trace amounts of vanadium at sub-ng/mL level with 4-(2-pyridylazo)resorcinol. Talanta, 42, 1057-1065.
[12] Fernández de Córdova, M.L., Ruiz Medina, A., & Molina Díaz, A. (1997). Solid phase spectrophotometric microdetermination of iron with ascorbic acid and ferrozine. Fresenius Journal of Analytical Chemistry, 357, 44-49.
[13] Ortega Barrales, P., Domínguez Vidal, A., Fernández de Córdova, M.L., & Molina Díaz, A. (2001). Simultaneous determination of thiamine and pyridoxine in pharmaceuticals by using a single flow-through biparameter sensor. Journal of Pharmaceutical and Biomedical Analysis, 25, 619-630.
[14] Valero-Navarro, A., Salinas-Castillo, A., Fernández-Sánchez, J.F., Segura-Carretero, A., Mallavia, R., & Fernández-Gutiérrez, A. (2009). The development of a MIP-optosensor for the detection of monoamine naphthalenes in drinking water. Biosensors and Bioelectronics, 24, 2305-2311.
[15] Guardia, L., Badia, R., & Díaz-García, M.E. (2007). Molecularly imprinted sol-gels for nafcillin determination in milk-based products. Journal of Agricultural and Food Chemistry, 55, 566-570.
[16] Ruzicka, J., & Scampavia, L. (1999). From flow injection to bead injection. Analytical Chemistry, 71, 257-263.
[17] Ruedas Rama, M.J., Ruiz Medina, A., & Molina Díaz, A. (2003). Bead injection spectroscopic flow-through renewable surface sensors with commercial flow cells as an alternative to reusable flow-through sensors. Analytica Chimica Acta, 482, 209-217.
[18] Ruzicka, J. (2000). Lab on-valve: Microflow analyzer based on sequential and bead injection. Analyst, 125, 1053-1060.
[19] Lavorante, A.F., Pires, C.K., & Reis, B.F. (2006). Multicommuted flow system employing pinch solenoid valves and micro-pumps. Spectrophotometric determination of paracetamol in pharmaceutical formulations. Journal of Pharmaceutical and Biomedical Analysis, 42, 423-429.

[20] Molina-Díaz, A., Ruiz-Medina, A., & Fernández-de Córdova, M.L. (1999). Determination of ascorbic acid by use of a flow-through solid phase UV spectrophotometric system. Fresenius Journal of Analytical Chemistry, 363, 92-97.

[21] Ayora Cañada, M.J., Pascual Reguera, M.I., Ruiz Medina, A., Fernández de Córdova, M.L., & Molina Díaz, A. (2000). Fast determination of paracetamol by using a very simple photometric flow-through sensing device. Journal of Pharmaceutical and Biomedical Analysis, 22, 59-66.

[22] Ruiz-Medina, A., Fernández-de Córdova, M.L., & Molina-Díaz, A. (1999). Flow injection-solid phase spectrofluorometric determination of pyridoxine in presence of group B-vitamins. Fresenius Journal of Analytical Chemistry, 363, 265-269.

[23] Gong, Z., & Zhang, Z. (1997). An optosensor for riboflavin with C18 silica gel as a substrate. Analytica Chimica Acta, 339, 161-165.

[24] Chen, D., de Castro, M.D.L., & Valcárcel, M. (1991). Fluorometric flow-through sensor for the determination of pyridoxal. Microchemical Journal, 44, 215-221.

[25] Llorent-Martínez, E.J., Ortega-Barrales, P., & Molina-Díaz, A. (2008). Sequential injection multi-optosensor based on a dual-luminescence system using two sensing zones: Application to multivitamin determination. Microchimica Acta, 162, 199-204.

[26] Ruiz-Medina, A., Ortega-Barrales, P., Fernández-de Córdova, M.L., & Molina-Díaz, A. (2002). Use of a continuous flow solid-phase spectroscopic sensor using two sensing zones: Determination of thiamine and ascorbic acid. Journal of AOAC International, 85, 369-374.

[27] Llorent-Martínez, E.J., García-Reyes, J.F., Ortega-Barrales, P., & Molina-Díaz, A. (2005). Flow-through fluorescence-based optosensor with on-line solid-phase separation for the simultaneous determination of a ternary pesticide mixture. Journal of AOAC International, 88, 860-865.

[28] López-Flores, J., Fernández-de Córdova, M.L., & Molina-Díaz, A. (2009). Simultaneous flow-injection solid-phase fluorometric determination of thiabendazole and metsulfuron methyl using photochemical derivatization. Analytical Sciences, 25, 681-686.

[29] Llorent-Martínez, E.J., Ortega-Barrales, P., & Molina-Díaz, A. (2006). Multi-commutated flow-through multi-optosensing: A tool for environmental analysis. Spectroscopy Letters, 39, 619-629.

[30] Fernández de Córdova, M.L., Ortega Barrales, P., Rodríguez Torné, G., & Molina Díaz, A. (2003). A flow injection sensor for simultaneous determination of sulfamethoxazole and trimethoprim by using Sephadex SP C-25 for continuous on-line separation and solid phase UV transduction. Journal of Pharmaceutical and Biomedical Analysis, 2003, 669-677.

[31] Picón-Zamora, D., Martínez-Galera, M., Garrido-Frenich, A., & Martínez-Vidal, J.L. (2000). Trace determination of carbendazim, fuberidazole and thiabendazole in water by application of multivariate calibration to cross-sections of 3-dimensional excitation- emission matrix fluorescence. Analyst, 125, 1167-1174.

[32] Ruiz Medina, A., Fernandez de Córdova, M.L., & Molina-Diaz, A. (1999). Simultaneous determination of paracetamol, caffeine and acetylsalicylic acid by means of a FI ultraviolet PLS multioptosensing device. Journal of Pharmaceutical and Biomedical Analysis, 21, 983-992.

[33] Ayora-Cañada, M.J., Pascual-Reguera, M.I., Molina-Díaz, A., & Capitán-Vallvey, L.F. (1999). Solid-phase UV spectroscopic multisensor for the simultaneous determination of caffeine, dimenhydrinate and acetaminophen by using partial least squares multicalibration. Talanta, 49, 691-701.

[34] Jerónimo, P.C.A., Araújo, A.N., & Montenegro, M.C.B.S. (2004). Development of a sol-gel optical sensor for analysis of zinc in pharmaceuticals. Sensors and Actuators B, 103, 169-177.

[35] Jerónimo, P.C.A., Araújo, A.N., Montenegro, M.C.B.S., Pasquini, C., & Raimundo, J. (2004). Direct determination of copper in urine using a sol-gel optical sensor coupled to a multicommutated flow system. Analytical and Bioanalytical Chemistry, 380, 108-114.

[36] Casado-Terrones, S., Fernández-Sánchez, J.F., Cañabate Díaz, B., Segura Carretero, A., & Fernández-Gutiérrez, A. (2005). A fluorescence optosensor for analyzing naphazoline in pharmaceutical preparations: Comparison with other sensors. Journal of Pharmaceutical and Biomedical Analysis, 38, 785-789.

[37] Gong, Z., & Zhang, Z. (1997). Determination of tetracyclines with a modified β-cyclodextrin based fluorosensor. Analytica Chimica Acta, 351, 205-210.

[38] González, G.P., Hernando, P.F., & Durand Alegría, J.S. (2008). Determination of digoxin in serum samples using a flow-through

fluorosensor based on a molecularly imprinted polymer. Biosensors and Bioelectronics, 23, 1754-1758.

[39] López-Flores, J., Fernández-de Córdova, M.L., & Molina-Díaz, A. (2005). Implementation of flow-through solid phase spectroscopic transduction with photochemically induced fluorescence: Determination of thiamine. Analytica Chimica Acta, 535, 161-168.

[40] López-Flores, J., Fernández de Córdova, M.L., & Molina Díaz, A. (2007). Flow-through optosensor combined with photochemically induced fluorescence for simultaneous determination of binary mixtures of sulfonamides in pharmaceuticals, milk and urine. Analytica Chimica Acta, 600, 164-171.

[41] López-Flores, J., Molina-Díaz, A., & Fernández-de Córdova, M.L. (2007). Development of a photochemically induced fluorescence-based optosensor for the determination of imidacloprid in peppers and environmental waters. Talanta, 72, 991-997.

[42] Cañabate-Díaz, B., Casado-Terrones, S., Segura-Carretero, A., Fernández, J.M.C., & Gutiérrez, A.F. (2004). Comparison of three different phosphorescent methodologies in solution for the analysis of naphazoline in pharmaceutical preparations. Analytical and Bioanalytical Chemistry, 379, 30-34.

[43] Aboul-Enein, H.Y., Stefan, R.I., Van Standen, J.F., Zhang, X.R., García-Campaña, A.M., & Baeyens, W.R.G. (2000). Recent developments and applications of chemiluminescence sensors. Critical Reviews in Analytical Chemistry, 30, 271-289.

[44] Zhang, Z., Zhang, S., & Zhang, X. (2005). Recent developments and applications of chemiluminescence sensors. Analytica Chimica Acta, 541, 37-47.

[45] Huang, Y., Zhang, C., Zhang, X., & Zhang, Z. (1999). A novel chemiluminescence flow-through sensor for the determination of analgin. Fresenius Journal of Analytical Chemistry, 365, 381-383.

[46] Fuwei, W., Jinghua, Y., Ping, D., & Shenguang, G. (2010). Molecular imprinting-chemiluminescence sensor for the determination of amoxicillin. Analytical Letters, 43, 1033-1045.

[47] Llorent-Martínez, E. J., Domínguez-Vidal, A., Ortega-Barrales, P., & Molina-Díaz, A. (2008). Fast determination of salicylic acid in pharmaceuticals by using a terbium-sensitized luminescent SIA Optosensor. Journal of Pharmaceutical Sciences, 97, 791-797.

[48] Alava-Moreno, F., Diaz-Garcia, M.E., & Sanz-Medel, A. (1993). Room temperature phosphorescence optosensor for tetracyclines. Analytica Chimica Acta, 281, 637-644.

[49] Alava-Moreno, F., Valencia-González, M.J., & Diaz-Garcia, M.E. (1998). Room temperature phosphorescence optosensor for anthracyclines. Analyst, 123, 151-154.

[50] Miró, M., Frenzel, W., Estela, J.M., & Cerdá, V. (2001). A novel flow-through disk-based solid-phase extraction diffuse reflectance optrode. Application to preconcentration and determination of trace levels of nitrite. Analyst, 126, 1740-1746.

[51] García, S., Albero, I., Ortuño, J.A., & Sanchez-Pedreño, C. (2003). Flow-through bulk optode for the spectrophotometric determination of perchlorate. Microchimica Acta, 143, 59-63.

[52] Xavier, M.P., Vallejo, B., Marazuela, M.D., Moreno-Bondi, M.C., Baldini, F., & Falai, A. (2000). Fiber optic monitoring of carbamate pesticides using porous glass with covalently bound chloropheno red. Biosensors and Bioelectronics, 14, 895-905.

[53] Garrigues, S., Vidal, M.T., Gallignani, M., & de la Guardia, M. (1994). On-line preconcentration and flow analysis-Fourier transform infrared determination of carbaryl. Analyst, 119, 659-664.

[54] Lendl, B., & Schindler, R. (1999). Flow-through sensors enhancing sensitivity and selectivity of FTIR spectroscopy in aqueous media. Vibrational Spectroscopy, 19, 1-10.

[55] Haberkorn, M., Hinsmann, P., & Lendl, B. (2002). A mid-IR flow-through sensor for direct monitoring of enzyme catalyzed reactions. Case study: Measurement of carbohydrates in beer. Analyst, 127, 109-113.

[56] Ortega-Barrales, P., Ayora-Cañada, M.J., Molina-Díaz, A., Garrigues, S., & de la Guardia, M. (1999). Solid phase Fourier transform near infrared spectroscopy. Analyst, 124, 579-582.

[57] Alcudia-León, M.C., Lucena, R., Cárdenas, S., & Valcárcel, M. (2008). Characterization of an attenuated total reflection-based sensor for integrated solid-phase extraction and infrared detection. Analytical Chemistry, 80, 1146-1151.

[58] Armenta, S., & Lendl, B. (2009). Flow through FTIR sensor based on solid phase spectroscopy (SPS) on conventional octadecyl (C18) silica. Vibrational Spectroscopy, 51, 60-64.

[59] Armenta, S., Garrigues, S., & de la Guardia, M. (2005). Solid-phase FT-Raman determination of caffeine in energy drinks. Analytica Chimica Acta, 547, 197-203.

[60] Ruedas Rama, M.J., López-Sánchez, M., Ruiz-Medina, A., Molina-Díaz, A., & Ayora-Cañada, M.J. (2005) Flow-through sensor with Fourier transform Raman detection for determination of sulfonamides. Analyst, 130, 1617-1623.

[61] Ortega Barrales, P., Fernández de Córdova, M.L., & Molina Díaz, A. (1998). A selective optosensor for UV spectrophotometric determination of thiamine in the presence of other vitamins B. Analytica Chimica Acta, 376, 227-233.

[62] Ortega-Barrales, P., Ruiz-Medina, A., Fernández-de Córdova, M.L., & Molina-Diaz, A. (1999). Sensitive and simple determination of diclofenac sodium by use of a continuous flow-injection procedure with solid-phase spectroscopic detection. Analytical Sciences, 15, 985-989.

[63] Ruiz-Medina, A., Fernández-de Córdova, M.L., & Molina-Díaz, A. (1999). Integrated flow injection-solid phase spectrophotometric determination of minoxidil. Talanta, 50, 277-282.

[64] Ruiz Medina, A., Fernández de Córdova, M.L., & Molina Díaz, A. (1999). Flow-through solid phase UV spectrophotometric determination of amoxycillin. Analytical Letters, 32, 729-742.

[65] Ayora Cañada, M.J., Pascual Reguera, M.I., & Molina Díaz, A. (2000). Selective determination of pyridoxine in the presence of hydrosoluble vitamins using a continuous-flow solid phase sensing device with UV detection. International Journal of Pharmaceutics, 202, 113-120.

[66] Ruiz-Medina, A., Fernández-de Córdova, M.L., & Molina-Díaz, A. (2000). Sensitive determination of adrenaline by means of a flow-through solid phase UV spectrophotometric sensing device. Microchimica Acta, 134, 101-105.

[67] Ruiz Medina, A., García Marín, M.G., Fernández de Córdova, M.L., & Molina Díaz, A. (2000). UV spectrophotometric flow-injection assay of tetracycline antibiotics retained on Sephadex QAE A-25 in drug formulations. Microchemical Journal, 65, 325-331.

[68] Ruiz-Medina, A., Fernández-de Córdova, M.L., Ortega-Barrales, P., & Molina-Díaz, A. (2001). Flow-through UV spectrophotometric

sensor for determination of (acetyl)salicylic acid in pharmaceutical preparations. International Journal of Pharmaceutics, 216, 95-104.

[69] Teixeira, L.S.G., Rocha, F.R.P., Korn, M., Reis, B.F., Ferreira, S.L.C., & Costa, A.C.S. (1999). Flow-injection solid-phase spectrophotometry for the determination of zinc in pharmaceutical preparations. Analytica Chimica Acta, 383, 309-315.

[70] Ruiz Medina, A., Cano García, M.C., & Molina Díaz, A. (2002). A flow analysis system for the rapid determination of sulfonamides using a solid phase photometric sensing zone. Analytical Letters, 35, 269-282.

[71] Ruedas-Rama, M.J., Ruiz Medina, A., & Molina-Díaz, A. (2003). New contributions to the field of bead-injection spectroscopy-flow-injection analysis: Determination of cobalt. Analytical and Bioanalytical Chemistry, 376, 527-533.

[72] Ruedas Rama, M.J., Ruiz Medina, A., & Molina Díaz, A. (2004). A Prussian blue-based flow-through renewable surface optosensor for analysis of ascorbic acid. Microchemical Journal, 78, 157-162.

[73] Pascual-Reguera, M.I., Pérez Parras, G., & Molina Díaz, A. (2004). A single spectroscopic flow-through sensing device for determination of ciprofloxacin. Journal of Pharmaceutical and Biomedical Analysis, 35, 689-695.

[74] Ruedas Rama, M.J., Ruiz Medina, A., & Molina Díaz, A. (2004). Bead injection spectroscopy-flow injection analysis (BIS-FIA): An interesting tool applicable to pharmaceutical analysis: Determination of promethazine and trifluoperazine. Journal of Pharmaceutical and Biomedical Analysis, 35, 1027-1034.

[75] Llorent-Martínez, E.J., García-Reyes, J.F., Ortega-Barrales, P., & Molina-Díaz, A. (2005). Solid-phase ultraviolet sensing system for determination of methylxanthines. Analytical and Bioanalytical Chemistry, 382, 158-163.

[76] Jerónimo, P.C.A., Araújo, A.N., Montenegro, M.C.B.S., Satinsky, D., & Solich, P. (2004). Colorimetric bismuth determination in pharmaceuticals using a xylenol orange sol-gel sensor coupled to a multicommutated flow system. Analytica Chimica Acta, 504, 235-241.

[77] Jerónimo, P.C.A., Araújo, A.N., Montenegro, M.C.B.S., Satinsky, D., & Solich, P. (2005). Flow-through sol-gel optical biosensor for the colorimetric determination of acetazolamide. Analyst, 130, 1190-1197.

[78] Ruiz-Medina, A., Fernández-de Córdova, M.L., Ayora-Cañada, M.J., Pascual-Reguera, M.I., & Molina-Díaz, A. (2000). A flow-through solid phase UV spectrophotometric biparameter sensor for the sequential determination of ascorbic acid and paracetamol. Analytica Chimica Acta, 404, 131-139.

[79] Ruiz Medina, A., Fernández de Córdova, M.L., & Molina Díaz, A. (1999). A very simple resolution of the mixture paracetamol and salicylamide by flow injection-solid phase spectrophotometry. Analytica Chimica Acta, 394, 149-158.

[80] Dominguez-Vidal, A., Garcia-Reyes, J.F., Ortega-Barrales, P., & Molina-Diaz, A. (2002). UV spectrophotometric flow-through multiparameter sensor for the simultaneous determination of acetaminophen, acetylsalicylic acid, and caffeine. Analytical Letters, 35, 2433-2447.

[81] Ortega-Barrales, P., Ruiz-Medina, A., Fernández-de Córdova, M.L., & Molina-Díaz, A. (2002). A flow-through solid-phase spectroscopic sensing device implemented with FIA solution measurements in the same flow cell: Determination of binary mixtures of thiamine with ascorbic acid or acetylsalicylic acid. Analytical and Bioanalytical Chemistry, 373, 227-232.

[82] Ortega-Barrales, P., Padilla-Weigand, R., & Molina-Díaz, A. (2002). Simultaneous determination of paracetamol and caffeine by flow injection-solid phase spectrometry using C18 silica gel as a sensing support. Analytical Sciences, 18, 1241-1246.

[83] Domínguez Vidal, A., Ortega Barrales, P., & Molina-Díaz, A. (2003). Simultaneous determination of paracetamol, caffeine and propyphenazone in pharmaceuticals by means of a single flow-through UV multiparameter sensor. Microchimica Acta, 141, 157-163.

[84] Ruedas-Rama, M.J., Ruiz-Medina, A., & Molina-Díaz, A. (2005). Resolution of biparametric mixtures using bead injection spectroscopic flow-through renewable surface sensors. Analytical Sciences, 21, 1079-1084.

[85] Llorent-Martínez, E.J., Domínguez-Vidal, A., Ortega-Barrales, P., de la Guardia, M., & Molina-Díaz, A. (2005). Implementation of multicommutation principle with flow-through multioptosensors. Analytica Chimica Acta, 545, 113-118.

[86] García-Reyes, J.F., Llorent-Martínez, E.J., Ortega-Barrales, P., & Molina-Díaz, A. (2006). The potential of combining solid-phase

optosensing and multicommutation principles for routine analyses of pharmaceuticals. Talanta, 68, 1482-1488.
[87] Gilbert-López, B., Llorent-Martínez, E.J., Ortega-Barrales, P., & Molina-Díaz, A. (2007). Development of a multicommuted flow-through optosensor for the determination of a ternary pharmaceutical mixture. Journal of Pharmaceutical and Biomedical Analysis, 43, 515-521.
[88] Ruiz-Medina, A., Fernández-de Córdova, M.L., & Molina-Díaz, A. (2001). A flow-through optosensing device with fluorometric transduction for rapid and sensitive determination of dipyridamole in pharmaceuticals and human plasma. European Journal of Pharmaceutical Sciences, 13, 385-391.
[89] Domínguez-Vidal, A., Ortega-Barrales, P., & Molina-Díaz, A. (2002). Fast flow-injection fluorometric determination of amiloride by using a solid sensing zone. Talanta, 56, 1005-1013.
[90] Fernández-Sánchez, J.F., Segura-Carretero, A., Cruces-Blanco, C., & Fernández-Gutiérrez, A. (2002). Room-temperature luminescence optosensings based on immobilized active principles actives: Application to nafronyl and naproxen determination in pharmaceutical preparations and biological fluids. Analytica Chimica Acta, 462, 217-224.
[91] Domínguez-Vidal, A., Ortega-Barrales, P., & Molina-Díaz, A. (2002). Determination of triamterene by transitory retention in a continuous flow solid phase system with fluorometric transduction. Journal of Pharmaceutical and Biomedical Analysis, 28, 721-728.
[92] Fernández-Sánchez, J.F., Carretero, A.S., Cruces-Blanco, C., & Fernández-Gutiérrez, A. (2003). A sensitive fluorescence optosensor for analyzing propranolol in pharmaceutical preparations and a test for its control in urine in sport. Journal of Pharmaceutical and Biomedical Analysis, 31, 859-865.
[93] Pascual-Reguera, M.I., Guardia-Rubio, M., & Molina-Díaz, A. (2004). Native fluorescence flow-through optosensor for the fast determination of diphenhydramine in pharmaceuticals. Analytical Sciences, 20, 799-803.
[94] Ortega-Algar, S., Ramos-Martos, N., & Molina-Díaz, A. (2004). Fluorometric flow-through sensing of quinine and quinidine. Microchimica Acta, 147, 211-217.
[95] Ruedas Rama, M.J., Ruiz Medina, A., & Molina Díaz, A. (2005). A flow-injection renewable surface sensor for the fluorometric

determination of vanadium(V) with Alizarin Red S. Talanta, 66, 1333-1339.

[96] López-Flores, J., Fernández-de Córdova, M.L., & Molina-Díaz, A. (2007). Determination of sub-ppb reserpine by an optosensing device based on photochemically induced fluorescence. Analytical and Bioanalytical Chemistry, 388, 1771-1777.

[97] González, G.P., Hernando, P.F., & Durand Alegría, J.S. (2009). A MIP-based flow-through fluoroimmunosensor as an alternative to immunosensors for the determination of digoxin in serum samples. Analytical and Bioanalytical Chemistry, 394, 963-970.

[98] Llorent-Martínez, E.J., Satinsky, D., & Solich, P. (2007). Fluorescence optosensing implemented with sequential injection analysis: A novel strategy for the determination of labetalol. Analytical and Bioanalytical Chemistry, 387, 2065-2069.

[99] Llorent-Martínez, E.J., Satinsky, D., Solich, P., Ortega-Barrales, P., & Molina-Díaz, A. (2007). Fluorometric SIA optosensing in pharmaceutical analysis: Determination of paracetamol. Journal of Pharmaceutical and Biomedical Analysis, 45, 318-321.

[100] López-Flores, J., Fernández-de Córdova, M.L., & Molina-Díaz, A. (2007). Multicommutated flow-through optosensors implemented with photochemically induced fluorescence: Determination of flufenamic acid. Analytical Biochemistry, 361, 280-286.

[101] Ruedas Rama, M.J., Ruiz Medina, A., & Molina Díaz, A. (2001). A flow-through sensing device with fluorometric transduction for the determination of warfarin by using an anion-exchanger gel combined with an FIA system. Analytical Sciences, 17, 1007-1010.

[102] Ruiz Medina, A., Fernández de Córdova, M.L., & Molina Díaz, A. (1999). A simple solid phase spectrofluorometric method combined with flow analysis for the rapid determination of salicylamide and salicylic acid in pharmaceutical samples. Fresenius Journal of Analytical Chemistry, 365, 619-624.

[103] Chen, D., de Castro, M.D.L., & Valcárcel, M. (1992). Flow-through sensing device based on derivative synchronous fluorescence measurements for the multi-determination of B6 vitamers. Analytica Chimica Acta, 261, 269-274.

[104] Llorent-Martínez, E.J., Ortega-Barrales, P., & Molina-Díaz, A. (2005). Multicommuted flow-through fluorescence optosensor for determination of furosemide and triamterene. Analytical and Bioanalytical Chemistry, 383, 797-803.

[105] Llorent-Martínez, E.J., García-Reyes, J.F., Ortega-Barrales, P., & Molina-Díaz, A. (2006). A multicommuted fluorescence-based sensing system for simultaneous determination of Vitamins B2 and B6. Analytica Chimica Acta, 555, 128-133.

[106] García-Reyes, J.F., Ortega-Barrales, P., & Molina-Díaz, A. (2007). Multicommuted fluorometric multiparameter sensor for simultaneous determination of naproxen and salicylic acid in biological fluids. Analytical Sciences, 23, 423-428.

[107] Cuenca-Trujillo, R.M., Ayora-Cañada, M., & Molina-Díaz, A. (2002). Determination of ciprofloxacin with a room-temperature phosphorescence flow-through sensor based on lanthanide-sensitized luminescence. Journal of AOAC International, 85, 1268-1272.

[108] Li, B., Zhang, Z., Zhao, L., & Xu, C. (2002). Chemiluminescence flow-through sensor for ofloxacin using solid-phase PbO_2 as an oxidant. Talanta, 57, 765-771.

[109] Li, B., Zhang, Z., Zhao, L., & Xu, C. (2002). Chemiluminescence flow-through sensor for pipemidic acid using solid sodium bismuthate as an oxidant. Analytica Chimica Acta, 459, 19-24.

[110] Song, Z.H., & Zhang, N. (2003). Chemiluminescence flow-through sensor for determination of reserpine in pharmaceutical preparation and biological fluids using immobilized reagents technology. Analytical Letters, 36, 41-57.

[111] Zhao, L., Li, B., Zhang, Z., & Lin, J.M. (2004). Chemiluminescent flow-through sensor for automated dissolution testing of analgin tablets using manganese dioxide as oxidate. Sensors and Actuators B, 97, 266-271.

[112] Llorent Martínez, E.J., García Reyes, J.F., Ortega Barrales, P., & Molina Díaz, A. (2005). Terbium-sensitized luminescence optosensor for the determination of norfloxacin in biological fluids. Analytica Chimica Acta, 532, 159-164.

[113] Zhou, H., Zhang, Z., He, D., & Xiong, Y. (2005). Flow through chemiluminescence sensor using molecularly imprinted polymer as recognition elements for detection of salbutamol. Sensors and Actuators B, 107, 798-804.

[114] Jeon, C.W., Khan, M.A., Lee, S.H., Karim, M.M., Lee, H.K., Suh, Y.S., Alam, S.M., & Chung, H.Y. (2008). Optical flow-through sensor for the determination of norfloxacin based on emission of $KMnO_4$-Na_2SO_3-Tb^{3+} system. Journal of Fluorescence, 18, 843-851.

References

[115] Ortega-Algar, S., Ramos-Martos, N., & Molina-Díaz, A. (2008). Flow-injection solid surface lanthanide-sensitized luminescence sensor for determination of p-aminobenzoic acid. Analytical and Bioanalytical Chemistry, 391, 715-719.

[116] Yu, J., Dai, P., Wan, F., Li, B., & Ge, S. (2009). Quantification of fenfluramine with a molecularly imprinted chemiluminescence sensor and sulfonophenylazo rhodanine. Journal of Separation Science, 32, 2170-2179.

[117] Molina-García, L., Llorent-Martínez, E.J., Fernández-de Córdova, M.L., & Ruiz-Medina, A. (2009). Development of a rapid and automatic optosensor for the determination of cromolyn in biological samples. Talanta, 79, 627-632.

[118] Llorent-Martínez, E.J., Ortega-Barrales, P., & Molina-Díaz, A. (2005). Multicommuted optosensor for the determination of pipemidic acid in biological fluids. Analytical Biochemistry, 347, 330-332.

[119] Llorent-Martínez, E.J., Ortega-Barrales, P., & Molina-Díaz, A. (2006). Chemiluminescence optosensing implemented with multicommutation: Determination of salicylic acid. Analytica Chimica Acta, 580, 149-154.

[120] Llorent-Martínez, E.J., Ortega-Barrales, P., Fernández de Córdova, M.L., & Ruiz-Medina, A. (2009). Development of an automated chemiluminescence flow-through sensor for the determination of 5-aminosalicylic acid in pharmaceuticals: A comparative study between sequential and multicommutated flow techniques. Analytical and Bioanalytical Chemistry, 394, 845-853.

[121] Molina-García, L., Llorent-Martínez, E.J., Fernández-de Córdova, M.L., & Ruiz-Medina, A. (2010). Direct determination of cefadroxil by chemiluminescence using a multicommutated flow-through sensor. Spectroscopy Letters, 43, 60-67.

[122] Llorent-Martínez, E.J., Ortega-Barrales, P., Molina-Díaz, A., & Ruiz-Medina, A. (2008). Implementation of terbium-sensitized luminescence in sequential-injection analysis for automatic analysis of orbifloxacin. Analytical and Bioanalytical Chemistry, 392, 1397-1403.

[123] Capitán-Vallvey, L.F., Valencia, M.C., & Arana Nicolás, E. (2003). Simple resolution of butylated hidroxyanisole and n-propyl gallate in fat foods and cosmetics samples by flow-injection solid phase spectrophotometry. Journal of Food Sciences, 68, 1595-1599.

[124] Capitán-Vallvey, L.F., Valencia, M.C., Nicolás, A.E., & García-Jiménez, J.F. (2006). Resolution of an intense sweetener mixture by use of a flow injection sensor with on-line solid-phase extraction. Analytical and Bioanalytical Chemistry, 385, 385-391.

[125] García-Reyes, J.F., Ortega-Barrales, P., & Molina-Díaz, A. (2005). Rapid determination of diphenylamine residues in apples and pears with a single multicommuted fluorometric optosensor. Journal of Agricultural and Food Chemistry, 53, 9874-9878.

[126] Molina-García, L., Ruiz-Medina, A., & Fernández-de Córdova, M.L. (2001). Automatic optosensing device based on photo-induced fluorescence for determination of piceid in cocoa-containing products. Analytical and Bioanalytical Chemistry, 399, 965–972.

[127] Molina-García, L., Ruiz-Medina, A., & Fernández-de Córdova, M.L. (2001). A novel multicommuted fluorometric optosensor for determination of resveratrol in beer, Talanta, 83, 850-856.

[128] Matsuoka, S., Nakatsu, Y., Takehara, K., Saputro, S., & Yoshimura, K. (2006). On-line electrochemical oxidation of Cr(III) using flow electrolysis cell for the determination of total Cr by flow injection-solid phase spectrophotomety. Analytical Sciences, 22, 1519-1524.

[129] Teixeira, L.S.G., & Rocha, F.R.P. (2007). A green analytical procedure for sensitive and selective determination of iron in water samples by flow injection solid-phase spestrophotometry. Talanta, 71, 1507-1511.

[130] Liu, Y.M., Garcia, R.P., Diaz Garcia, M.E., & Sanz-Medel, A. (1991). Room-temperature luminescence optosensing based on immobilized metal chelates: Application to iodide determination. Analytica Chimica Acta, 255, 245-251.

[131] San Vicente de la Riva, B., Costa-Fernández, J.M., Pereiro, & R., Sanz-Medel, A. (1999). Flow-through room temperature phosphorescence optosensing for the determination of lead in sea water. Analytica Chimica Acta, 395, 1-9.

[132] Segura-Carretero, A., Costa-Fernández, J.M., Pereiro, R., & Sanz-Medel, A. (1999). Low-level mercury determination with thiamine by fluorescence optosensing. Talanta, 49, 907-913.

[133] Gong, Z., & Zhang, Z. (1996). An optosensor based on the fluorescence of metal complexes adsorbed on Chelex 100. Analytica Chimica Acta, 325, 201-204.

[134] García-Reyes, J.F., Ortega-Barrales, P., & Molina-Díaz, A. (2007). Flow-through fluorescence-based optosensor for the screening of zinc in drinking water. Analytical Sciences, 23, 1179-1183.

[135] García-Reyes, J.F., Ortega-Barrales, P., & Molina-Díaz, A. (2006). Sensing of trace amounts of cadmium in drinking water using a single fluorescence-based optosensor. Microchemical Journal, 82, 94-99.

[136] García-Reyes, J.F., Ortega-Barrales, P., & Molina-Díaz, A. (2005). Development of a solid surface fluorescence-based sensing system for aluminium monitoring in drinking water. Talanta, 65, 1203-1208.

[137] Ruedas Rama, M.J., Ruiz Medina, A., & Molina Díaz, A. (2004). Implementation of flow-through multisensors with bead injection spectroscopy: Fluorometric renewable surface biparameter sensor for determination of berillium and aluminium. Talanta, 62, 879-886.

[138] Fernández-Band, B., Linares, P., Luque de Castro, M.D., & Valcárcel, M. (1991). Flow through sensor for the direct determination of pesticide mixtures without chromatographic separation. Analytical Chemistry, 63, 1672-1675.

[139] Guzmán Mar, J.L., López Martínez, L., López de Alba, P.L., Castrejón Durán, J.E., & Cerdá Martín, V. (2006). Optical fiber reflectance sensor coupled to a multisyringe flow injection system for preconcentration and determination of 1-naphthylamine in water samples. Analytica Chimica Acta, 573-574, 406-412.

[140] Agudo, M., Rios, A., & Valcarcel, M. (1993). Automatic continuous-flow determination of paraquat at the subnanogram per millilitre level Analytica Chimica Acta, 281, 103-109.

[141] Anastassiades, M., Lehotay, S.J., Stajnbaher, D., & Schenck, F.J. (2003). Fast and easy multiresidue method employing acetonitrile extraction/partitioning and "dispersive solid-phase extraction" for the determination of pesticide residues in produce. Journal of AOAC International, 86, 412-431.

[142] González-Martínez, M.A., Maquieira, A., & Puchades, R. (2003). International Journal of Environmental Analytical Chemistry, 83, 633-642.

[143] Podola, B., & Melkonian, M. (2005). Selective real-time herbicide monitoring by an array chip biosensor employing diverse microalgae. Journal of Applied Phycology, 17, 261-271.

[144] Penalva, J., Gabaldon, J.A., Maquieira, A., & Puchades, R. (2000). Determination of carbaryl in vegetables using an immunosensor

working in organic media. Food and Agricultural Immunology, 12, 101-114.

[145] González-Martínez, M.A., Morais, S., Puchades, R., Maquieira, A., Abad, A., & Montoya, A. (1997). Development of an automated controlled-pore glass flow-through immunosensor for carbaryl. Analytica Chimica Acta, 347, 199-205.

[146] Pulido-Tofiño, P., Barrero-Moreno, J.M., & Pérez-Conde, M.C. (2000). Flow-through fluoroimmunosensor for isoproturon determination in agricultural foodstuff. Evaluation of antibody immobilization on solid support. Analytica Chimica Acta, 417, 85-94.

[147] Penalva, J., Puchades, R., Maquieira, A., Gee, S., & Hammock, B.D. (2000). Development of immunosensors for the analysis of l-naphthol in organic media. Biosensors and Bioelectronics, 15, 99-106.

[148] López-Flores, J., Molina-Díaz, A., & Fernández-de Córdova, M.L. (2007). Determination of azoxystrobin residues in grapes, musts and wines with a multicommuted flow-through optosensor implemented with photochemically induced fluorescence. Analytica Chimica Acta, 585, 185-191.

[149] García-Reyes, J.F., Ortega-Barrales, P., & Molina-Díaz, A. (2004). Development of a single fluorescence-based optosensor for rapid simultaneous determination of fungicides benomyl and thiabendazole in waters and commercial formulations. Journal of Agricultural and Food Chemistry, 52, 2197-2202.

[150] Llorent-Martínez, E.J., García-Reyes, J.F., Ortega-Barrales, P., & Molina-Díaz, A. (2007). Multicommuted fluorescence based optosensor for the screening of bitertanol residues in banana samples. Food Chemistry, 102, 676-682.

[151] Sánchez-Barragán, I., Karim, K., Costa-Fernández, J.M., Piletsky, S.A., & Sanz-Medel, A. (2007). A molecularly imprinted polymer for carbaryl determination in water. Sensors and Actuators B, 123, 798-804.

[152] García-Reyes, J.F., Llorent-Martínez, E.J., Ortega-Barrales, P., & Molina-Díaz, A. (2004). Multiwavelength fluorescence based optosensor for simultaneous determination of fuberidazole, carbaryl and benomyl. Talanta, 64, 742-749.

[153] Piccirilli, G.N., Escandar, G.M., Cañada, F.C., Merás, I.D., & Muñoz de la Peña, A. (2008). Flow-through photochemically induced fluorescence optosensor for the determination of linuron. Talanta, 77, 852-857.

[154] López-Flores, J., Fernández-de Córdova, M.L., & Molina-Díaz, A. (2009). Flow-through optosensing device implemented with photochemically-induced fluorescence for the rapid and simple screening of metsulfuron methyl in environmental waters. Journal of Environmental Monitoring, 11, 1080-1085.

[155] Domínguez-Vidal, A., Ortega-Barrales, P., & Molina-Díaz, A. (2007). Environmental water samples analysis of pesticides by means of chemometrics combined with fluorometric multioptosensing. Journal of Fluorescence, 17, 271-277.

[156] Herranz, S., Ramón-Azcón, J., Benito-Peña, E., Marazuela, M.D., Marco, M.P., & Moreno-Bondi, M.C. (2008). Preparation of antibodies and development of a sensitive immunoassay with fluorescence detection for triazine herbicides. Analytical and Bioanalytical Chemistry, 391, 1801-1812.

[157] García-Reyes, J.F., Llorent-Martínez, E.J., Ortega-Barrales, P., & Molina-Díaz, A. (2006). Determination of thiabendazole residues in citrus fruits using a Multicommuted fluorescence-based optosensor. Analytica Chimica Acta, 557, 95-100.

[158] Ruedas-Rama, M.J., Ruiz-Medina, A., & Molina-Díaz, A. (2002). Use of a solid sensing zone implemented with unsegmented flow analysis for simultaneous determination of thiabendazole and warfarin. Analytica Chimica Acta, 459, 235-243.

[159] Badía, R., & Díaz-García, M.E. (1999). Cyclodextrin-based optosensor for the determination of warfarin in waters. Journal of Agricultural and Food Chemistry, 47, 4256-4260.

[160] Roda, A., Rauch, P., Ferri, E., Girotti, S., Ghini, S., Carrea, G., & Bovara, R. (1994). Chemiluminescent flow sensor for the determination of paraoxon and aldicarb pesticides. Analytica Chimica Acta, 294, 35-42.

[161] Varsamis, D.G., Touloupakis, E., Morlacchi, P., Ghanotakis, D.F., Giardi, M.T., & Cullen, D.C. (2008). Development of a photosystem II-based optical microfluidic sensor for herbicide detection. Talanta, 77, 42-47.

[162] Park, I.S., & Kim, N. (2006). Development of a chemiluminescent immunosensor for chloramphenicol. Analytica Chimica Acta, 578, 19-24.

[163] Surugiu, I., Svitel, J., Ye, L., Haupt, K., & Danielsson, B. (2001). Development of a flow-injection capillary chemiluminescent ELISA

using an imprinted polymer instead of the antibody. Analytical Chemistry, 73, 4388-4392.

[164] Ayyagari, M.S., Kamtekar, S., Pande, R., Marx, K.A., Kumar, J., Tripathy, S.K., & Kaplan, D.L. (1995). Aspects of a biosensor development. Materials Science & Engineering, C2, 191-196.

[165] Fang, Y., Yan, S., Ning, B., Liu, N., Gao, Z., & Chao, F. (2009). Flow injection chemiluminescence sensor using molecularly imprinted polymers as recognition element for determination of maleic hydrazide. Biosensors and Bioelectronics, 24, 2323-2327.

[166] Salinas-Castillo, A., Fernández-Sánchez, J.F., Segura-Carretero, A., & Fernández-Gutiérrez, A. (2004). A facile flow-through phosphorescence sensing device for simultaneous determination of naptalam and its metabolite 1-naphthylamine. Analytica Chimica Acta, 522, 19-24.

[167] Piccirilli, G.N., & Escandar, G.M. (2009). Flow injection analysis with on-line nylon powder extraction for room-temperature phosphorescence determination of thiabendazole. Analytica Chimica Acta, 646, 90-96.

[168] Fenoll, J., Hellín, P., Martínez, C.M., & Flores, P. (2009). Multiresidue analysis of pesticides in soil by high-performance liquid chromatography with tandem mass spectrometry. Journal of AOAC International, 92, 1566-1575.

INDEX

A

acetaminophen, 56, 102, 107
acetic acid, 67, 85, 86
acetonitrile, 92, 113
acetylcholinesterase, 67, 92
acid, 13, 51, 52, 53, 56, 65, 71, 72, 73, 74, 75, 76, 77, 78, 79, 80, 81, 84, 85, 86, 87, 90, 91, 95, 100, 101, 102, 103, 106, 107, 109, 110, 111
acidic, 63, 79
adsorption, 17, 67, 88
aluminium, 60, 90, 91, 113
amiloride, 75, 108
amino acids, 63
antibody, 114, 116
aqueous solutions, 15, 62, 68
ascorbic acid, 51, 53, 71, 78, 100, 101, 106, 107
aspiration, 42, 43, 45, 49
atoms, 61
automation, vii, 2, 3, 21, 27, 48, 49, 92, 98

B

background noise, 12, 56
beer, 68, 89, 104, 112
beryllium, 52, 91

biological fluids, 7, 60, 71, 75, 98, 108, 110, 111
biological samples, 71, 97, 111
biomolecules, 13
biosensors, 31, 62
biosphere, 89
bismuth, 71, 106
Brazil, 43, 46

C

cadmium, 113
caffeine, 56, 68, 69, 75, 102, 105, 107
carbohydrates, 68, 104
catalyst, 13, 63
cation, 12, 63, 82, 86
chelates, 63, 65, 112
chemical, 1, 3, 12, 23, 26, 28, 31, 52, 59, 61, 62, 63, 66, 71, 87, 97, 99
chemiluminescence, 7, 30, 103, 110, 111, 116
chemometrics, 56, 92, 115
chromatography, 21, 97
chromium, 89, 99
CL sensors, 62, 63
cobalt, 99, 106
cocoa, 88, 112
commercial, 19, 24, 25, 54, 59, 68, 100, 114
compaction, 18, 26

compounds, 27, 56, 58, 59, 62, 65, 71, 78, 89, 97
computer, 20, 39, 42, 43
configuration, vii, 15, 17, 19, 23, 34, 46, 60
consumption, vii, 2, 9, 27, 43, 45, 49, 97
contamination, 18, 86
copper, 71, 99, 102
cost, vii, 10, 19, 46, 58, 62, 97, 98
cyclodextrins, 61, 64

D

decay, 61, 63, 65
degradation, 28, 91
detection techniques, vii, 31, 58, 78
diffusion, 41, 63, 65
dimethacrylate, 63
diphenhydramine, 108
dispersion, 31, 33, 34, 38
distribution, 6, 13, 18, 23, 24, 27
drinking water, 100, 113

E

ELISA, 115
emission, 1, 12, 59, 61, 62, 63, 64, 65, 82, 83, 102, 110
emission wavelengths, 63, 64
energy, 61, 63, 64, 65, 69, 83, 105
enzyme, 62, 63, 67, 104
ethanol, 13, 88
ethylene glycol, 63
European Union (EU), 92
europium-sensitized luminescence, 66
excitation, 6, 59, 61, 63, 64, 65, 82, 83, 102
extraction, 7, 21, 67, 68, 88, 89, 91, 92, 104, 112, 113, 116

F

fiber, 57, 66, 113
fibroblasts, 18

Flow Injection-Solid Phase Spectroscopy (FI-SPS), 2
Flow Through Optosensors (FTO), 2
fluorescence, 6, 7, 14, 25, 28, 30, 31, 52, 54, 58, 59, 60, 61, 71, 78, 81, 91, 99, 101, 102, 103, 108, 109, 110, 112, 113, 114, 115
fluorescent compounds, 78
fluorophores, 59
formation, 14, 27, 31, 42, 52, 65, 68, 83, 86, 89
FTIR, 67, 104

G

gel, 11, 13, 18, 51, 52, 54, 56, 62, 67, 68, 69, 73, 76, 78, 92, 96, 101, 107, 109
growth, 97

H

herbicide, 113, 115
homogeneity, 18, 23
hydrogen, 14, 62, 67
hydrolysis, 13, 37, 67, 72, 91

I

immobilization, 7, 15, 28, 61, 62, 96, 114
imprinting, 63, 103
infrared spectroscopy, 67
inhibition, 92, 95
insertion, 31, 32, 42, 44, 81
interference, 13, 14, 56, 67, 68, 69, 92
ion-exchange, 1, 7, 11, 26, 53, 57, 99
ions, 63, 64, 65, 67, 83
IR transmission, 68
IRC, 79
iron, 89, 99, 100, 112
irradiation, 28, 60, 62

Index

K

ketones, 65
kinetics, 53

L

laminar, 41
lanthanide, 25, 63, 64, 65, 66, 82, 110, 111
lifetime, 11, 15, 28, 36, 62, 65
light, 1, 6, 7, 15, 23, 24, 25, 26, 57, 58, 59, 61, 62, 63, 64, 66, 83, 87, 88
liquid chromatography, 116
liquids, 43, 45, 46
luminescence, vi, 12, 25, 58, 61, 63, 64, 66, 67, 79, 81, 83, 95, 101, 108, 110, 111, 112

M

maltose, 68
manganese, 110
manifolds, 35, 42, 45
manipulation, 1, 8, 18, 21
Marx, 116
mass, 6, 7, 23, 31, 97, 116
mass spectrometry, 31, 97, 116
matrix, 11, 13, 17, 69, 102
media, 27, 64, 65, 104, 114
membranes, 28, 45, 67
mercury, 60, 90, 112
metal complexes, 112
metal ion, 59
metals, 27, 91
methacrylic acid, 63
methanol, 13, 54, 55, 56, 67, 68, 89
methodology, vii, 1, 2, 3, 7, 10, 17, 19, 42, 45, 65, 71, 84, 89, 91, 92
micro-pumps, 43, 46, 100
miniaturization, 3, 9, 27
minicolumn, 53, 54, 55, 74, 94
MIP, 14, 59, 63, 76, 79, 80, 81, 88, 93, 95, 96, 100, 109

mixing, 13, 35, 42, 45, 46, 85
molecules, 13, 21, 52, 61, 65, 83
Montenegro, 102, 106
Multicommuted Flow Analysis (MCFIA), 42
multicommuted flow techniques, 3, 42
Multipumping Flow Systems (MPFS), 42
Multisyringe Flow Analysis (MSFIA), 42
multivariate calibration, 56, 102

N

Na_2SO_4, 80
NaCl, 51, 52
naphthalene, 14, 91
near infrared spectroscopy, 104
NIR, 68
NIR spectra, 69
nitrite, 67, 104
nitrogen, 61
non-polar, 7, 11, 12, 27

O

ofloxacin, 81, 110
optical fiber, 17, 21, 57, 58, 66, 67, 91
organic compounds, 3, 7, 52, 59, 97
oxidation, 71, 81, 89, 112
oxygen, 61

P

parallel, 47
partial least-squares, 56
perchlorate, 67, 104
pesticide, 54, 60, 91, 101, 113
pH, 27, 28, 51, 52, 62, 65, 67, 73, 77, 83, 85, 86, 92
pharmaceutical, vii, 7, 10, 53, 56, 58, 59, 60, 66, 69, 71, 75, 81, 82, 84, 85, 93, 97, 99, 100, 102, 103, 106, 107, 108, 109, 110, 111
phosphate, 52, 77

phosphorescence, 7, 25, 58, 61, 65, 82, 104, 110, 112, 116
photodegradation, 59
photosynthesis, 93
PLS, 56, 73, 102
polycyclic aromatic hydrocarbon, 59
polymer, 11, 12, 13, 14, 63, 68, 103, 110, 114, 116
polysaccharide, 12
polystyrene, 57, 68
Portugal, 46
potassium, 79, 81, 95, 96
principles, 38, 69, 108
proportionality, 24
propranolol, 108
PTFE, 60, 66
PVC, 67
pyridoxine, 52, 75, 100, 101, 105

Q

quality control, 27
quantification, 91, 98
quartz, 57

R

radiation, 2, 28, 57, 64, 66
radical polymerization, 13
Raman spectroscopy, 69
reactants, 61
reaction medium, 63
reaction rate, 59
reaction time, 42
reaction zone, 62
reactions, 13, 59, 62, 104
reactive groups, 21
reagents, 2, 3, 21, 27, 28, 31, 39, 45, 46, 47, 49, 55, 59, 62, 81, 96, 98, 110
recognition, 13, 63, 96, 110, 116
regeneration, 9, 10, 15, 17, 32, 96
residues, 112, 113, 114, 115
resins, 7, 11, 12, 26
resistance, 29

resorcinol, 100
resveratrol, 89, 112
riboflavin, 52, 75, 101
room temperature, 13, 61, 112

S

saccharin, 88
scatter, 62
scattering, 28
selectivity, vii, 1, 7, 9, 10, 12, 13, 14, 27, 58, 62, 64, 68, 69, 82, 97, 98, 104
sensing, 6, 9, 10, 15, 17, 35, 36, 51, 52, 53, 54, 55, 60, 63, 75, 78, 84, 86, 101, 105, 106, 107, 108, 109, 110, 113, 115, 116
sensitivity, vii, 1, 6, 7, 23, 25, 48, 49, 54, 58, 59, 62, 64, 65, 67, 68, 82, 85, 89, 97, 98, 104
sensors, vii, 5, 6, 7, 10, 12, 15, 31, 36, 52, 55, 62, 63, 67, 71, 72, 75, 98, 99, 100, 102, 103, 104, 107
serum, 102, 109
signals, 56, 91, 96
signal-to-noise ratio, 24
silica, 12, 18, 52, 54, 56, 67, 68, 78, 88, 92, 96, 101, 104, 107
sodium, 19, 61, 81, 83, 85, 86, 95, 105, 110
solenoid coil, 45
solenoid piston pumps, 46
sol-gel, 11, 13, 62, 71, 100, 102, 106
solid phase, 1, 6, 8, 15, 17, 23, 57, 58, 59, 61, 65, 66, 68, 69, 100, 101, 102, 103, 104, 105, 106, 107, 108, 109, 111, 112
Solid Phase Spectroscopy (SPS), 1
solid state, 62
sorption, vii, 7, 13, 18, 23
Spain, 47
species, 1, 2, 3, 6, 7, 10, 11, 13, 14, 15, 17, 23, 24, 25, 27, 28, 31, 36, 42, 57, 58, 60, 61, 62, 65, 66, 68, 69, 71, 98
specific surface, 67
spectrophotometry, 90, 106, 107, 111
spectroscopic characteristics, 78
spectroscopy, 68, 71, 99, 104, 106, 113
spin, 65

sponge, 79, 81
stability, 14, 15, 28, 62, 81
standard deviation, 49
state, 27, 61, 62, 63, 64, 65, 89
strong interaction, 15
structure, 12, 13, 64
styrene polymers, 11
substrate, 62, 96, 101
sulfonamides, 60, 69, 103, 105, 106
sulfur dioxide, 83
surfactants, 65
synthesis, 13
synthetic polymers, 13

T

teflon, 29, 45
temperature, 13, 61, 104, 108, 110, 112, 116
template molecules, 14
template polymerization technique, 13
terbium, 65, 82, 83, 85, 86, 103, 111
tetracycline antibiotics, 105
tetracyclines, 59, 102, 104
trace elements, 71, 89, 90
transducer, 58, 92
transduction, 67, 102, 103, 108, 109
transport, 53
treatment, 3, 55, 56, 72, 76, 77, 79, 88, 90, 96

U

uric acid, 63
urine, 58, 60, 63, 66, 81, 85, 86, 102, 103, 108

UV, 11, 12, 24, 52, 54, 56, 57, 58, 59, 60, 71, 88, 90, 101, 102, 105, 107
UV irradiation, 59, 60
UV light, 60, 88
UV-irradiation, 52, 54
UV-Visible spectroscopy, 58

V

Valencia, 104, 111, 112
valve, 19, 20, 21, 32, 33, 34, 35, 36, 37, 39, 40, 41, 42, 43, 44, 45, 46, 47, 53, 83, 86, 87, 100
vanadium, 89, 100, 109
variables, 13, 39
versatility, 31, 42, 77
vitamin C, 52
vitamins, 52, 53, 71, 77, 78, 101, 105

W

waste, vii, 9, 20, 32, 33, 35, 41, 43, 98
water, 13, 14, 54, 67, 68, 69, 84, 89, 99, 102, 112, 113, 114, 115
wavelengths, 5, 63, 64, 91
windows, 68
working conditions, 53

Y

yield, 31

Z

zinc, 71, 102, 106, 113